AI and the Law

A Practical Guide to Using Artificial Intelligence Safely

Harry Borovick

Apress®

AI and the Law: A Practical Guide to Using Artificial Intelligence Safely

Harry Borovick
LONDON, UK

ISBN-13 (pbk): 979-8-8688-0399-4   ISBN-13 (electronic): 979-8-8688-0400-7
https://doi.org/10.1007/979-8-8688-0400-7

Managing Director, Apress Media LLC: Welmoed Spahr
Acquisitions Editor: Shivangi Ramachandran
Development Editor: James Markham
Editorial Project Manager: Jessica Vakili

Cover designed by eStudioCalamar

Distributed to the book trade worldwide by Springer Science+Business Media New York, 1 New York Plaza, Suite 4600, New York, NY 10004-1562, USA. Phone 1-800-SPRINGER, fax (201) 348-4505, e-mail orders-ny@springer-sbm.com, or visit www.springeronline.com. Apress Media, LLC is a California LLC and the sole member (owner) is Springer Science + Business Media Finance Inc (SSBM Finance Inc). SSBM Finance Inc is a **Delaware** corporation.

For information on translations, please e-mail booktranslations@springernature.com; for reprint, paperback, or audio rights, please e-mail bookpermissions@springernature.com.

Apress titles may be purchased in bulk for academic, corporate, or promotional use. eBook versions and licenses are also available for most titles. For more information, reference our Print and eBook Bulk Sales web page at http://www.apress.com/bulk-sales.

Any source code or other supplementary material referenced by the author in this book is available to readers on GitHub. For more detailed information, please visit https://www.apress.com/gp/services/source-code.

If disposing of this product, please recycle the paper

To Katie (and Arnie), always.

To my parents, Martyn and Corinne, for teaching me to work hard and giving me all the support and trust that I could have ever asked for.

Table of Contents

About the Author

Harry Borovick is General Counsel at Luminance, a company that provides advanced AI to process and negotiate legal documents. Previously, Harry worked to solve legal issues in complex regulated technologies, including gaming, FinTech and AdTech.

As well as working for a company at the forefront of the development of AI for streamlining legal processes, Harry has lectured at the University of Law, London, on contract law, BARBI on commercial law, and Queen Mary University of London on legal technology and AI. Harry also advises various organizations and legal bodies, including the Chartered Institute of Arbitrators. He is also a professional speaker on the subject of AI and data privacy, having spoken for numerous organizations, including media organizations, and has been invited to speak at events such as the Economist GC Summit.

About the Technical Reviewer

 **Connagh McCormick** is a barrister and experienced general counsel in building and leading high-performing legal teams in fast-growing businesses. A passionate advocate for the integration of AI in the legal industry, Connagh has been an early adopter of cutting-edge AI solutions, positioning himself at the forefront of driving change and advancing legal operations through technology in the evolving legal landscape.

Connagh frequently speaks at conferences to legal industry leaders and general counsels, sharing insights on how to leverage AI to improve efficiency and provide greater business support. His expertise includes guiding companies on the implementation of technology and change management strategies to modernize legal departments and enhance their strategic value to the broader business.

Acknowledgments

Thank you to the following kind contributors:

- Robert Webb K.C., as a mentor and for his generous foreword.

- David Hancock, for mentorship and guidance as well as his thoughts on AI in finance.

- Dr. Tim Battcock, for the engaging debates on what it means to be a professional and the importance of human-to-human interface.

- Jessica Hare, for guiding me to win the bet.

- Dr. George Williams, for his understanding of human kindness in modern medicine.

- My colleagues and friends at Luminance, for encouraging and supporting me during the writing of this book. Special thank you for the guidance and faith, Mike Lynch.

Foreword

Harry Borovick is well qualified to write this timely and authoritative book. As General Counsel of Luminance, he sees today's operation of legal tech at the highest and most relevant levels both in the corporate and professional spheres . He understands its possibilities but also is realistic about its limits.

Empirical research, due diligence, precedent and standard form contracts are in the top suite of AI's strengths: they are both necessary in today's data-driven world and growing fast. Analysis, reasoning, mercy, and compassion are weaker elements. Probability-based tools can be reluctant to embrace the unlikely or to push against the grain. They will not necessarily spot where reform is needed and may be tempted to follow precedent even where the past is a bad guide to the future. Much of the law thrives on ambiguity; much of AI prospers on probability. They are not the same thing and that interface is one of the hot topics of our time.

The story of AI is in its infancy. Harry's book makes a welcome introduction.

Robert Webb, KC

(Former General Counsel of Rolls-Royce and British Airways, and former Non-Executive Chairman of BBC Worldwide)

Introduction

Who I Am

I trained as a lawyer, like most, in a law firm. It was a great firm with great people. It did not have good tech. It is now one of the firms with the best tech, but I wasn't there to see it. All of that post-me progress happened in a very urgent fashion with the arrival of a few visionary and driven people. Before becoming a lawyer, I helped run and eventually sell my family business. This was an institution in our Soho (London), operating in the same building since 1932. But, when my father was unexpectedly and suddenly feeling too old to give the business his all, we needed to make a sudden change.

These experiences framed my perceptions and gave me the first clear realization of my professional life – the world moves swiftly and sometimes all at once. Being able to adapt to these sudden changes, with "North star" guiding principles, is what makes for long-lasting success.

I'm now a commercial and regulatory lawyer, leading the legal team of an AI company. I find myself here by working (mostly by happy accident) in a variety of industries which heavily rely on types of AI (in some cases it was very, very basic AI). Of course, there has been a rapid explosion of the term AI in our daily language and actual use of AI by ordinary people since 2022 with the release of ChatGPT. This has meant that I've found myself in a fortuitous position for two reasons:

1. Not many lawyers worked with anything resembling what we would now generally call AI prior to 2022.

2. Most of those lawyers who worked in AI-related tech prior to 2022 don't find themselves currently being able to work in such a way that they can fully use that experience in day-to-day work.

Lucky me. What this means though is that I've had to learn on the fly in a frustratingly fast-evolving field. One of my core goals in my day job is to help others learn the fundamentals of AI, to understand how practically useful it can be. This book aims to do the same. I've taken to professional lecturing with various universities and organizations, but the aim of this book is to condense and simplify some complex problems for practical day-to-day use. Anyone can explore these issues in greater depth should this book spark a deeper interest. At the back of this book, I have put together a summary which includes a list of suggested additional reading/ listening/watching.

What We Might Mean When We Say AI

Artificial Intelligence (AI), Artificial General Intelligence (AGI), Machine Learning (ML), Algorithms, Transformers, Generative Pre-trained Transformers (GPT), Generative Artificial Intelligence (GenAI)... the list goes on. The AI space is full of potentially confusing and often misused terms, some of which focus on outcomes (e.g., our perception by a user of a system seeming intelligent) while others focus on the process happening behind the scenes.

None of this is relevant to the scope of this book. What is relevant is that these terms – some of which are old, some of which are new – will continually evolve and adapt. Just like all technical and ordinary language, as the state-of-the-art or the cultural zeitgeist shifts, so does the meaning of existing terms and the emergence of new terms. So, the key is not to focus on the words which describe what is happening, whether before our eyes or *in the magic*. Instead, for 99% of us, we should focus on our own possible uses, benefits, and pitfalls of engaging with AI-related or AI-based technologies.

Really, what matters is outcomes. Outcomes, if done well, is our perception that a human-like (sometimes better than or faster than human) process, discussion, learning, or interaction is occurring. So, as that is close to one of the classic definitions of AI, for simplicity I aim to be mostly referring to "AI" as a bit of a catch-all. In some chapters, it may make sense to contextualize by touching on some processes, technicalities, and terms (for those who might be interested). However, the goal of this book – above all else – is accessibility. If it gets too technical, I've done a bad job (I look forward to the editorial complaints).

What I Hope to Share

As I write this book in late 2023, the market cap of global cryptocurrency remains north of US$1 trillion. This is despite the value of Bitcoin, the most well-known and inherently valuable/scarce token, being at less than half its peak value. What does this tell us about emerging technologies in the post-Internet-adoption era?

Firstly, that everyone remains in search of the next big hype bubble – "the biggest thing since the Internet" – crypto wasn't the first, it won't be the last. Secondly, that even the most bullish and experienced backers of an emerging and legitimately useful technology can get caught blind, overexcited, and lacking caution – we only have to look honestly in hindsight at the collapse of the once golden-child Sam Bankman-Fried and FTX. Finally, that the real skill in predicting the future of technology is understanding friction.

If a technology, like the Internet, fundamentally changes how ordinary people work or go about simple tasks and questions their lives and this aligns with the objectives of larger organizations (greater efficiencies and potentially lower costs), then it succeeds. Success is a low level of friction for initial adoption, ongoing use (by a wide variety of users) and progressively developing use cases as a result of the low-friction ecosystem.

Complexity arises when a fundamentally useful technology which is primed for mass adoption gets muddled, confused, or outright mis-sold to potential users. Usage of the Internet has passed the level where bad actors (con artists, charlatans, or even just over-enthusiastic and well-intentioned fools) prevent its mass-market daily usage.

Usage of AI has seen a meteoric rise in a comparatively short time. What is unique to the adoption of AI is the initial focus by the vast majority of users, in a personal and professional capacity, has been on value generation through efficiency. Whether that is a Maid of Honor using ChatGPT to help them write a good wedding speech, or a lawyer using a sophisticated tool to swiftly review large volumes of documents.

This isn't to say that, unlike the early stages of the Internet (and later the Crypto-boom), there isn't the opportunity to "get rich quick" for some. But, the success of the Internet has been based on long-term stable adoption and practical usage.

This isn't one of those books where I tell you some scheme to get wealthy quickly using a new technology. It isn't a book where I promise you will be able to out-compete the world with new tricks. The goals (and hopefully your outcomes) of this book are to realize that AI isn't coming, it's here – and it has opportunities and challenges. The key to navigating the early stages of the AI era is to understand guiding principles and then to layer the need-specific (whether professional or personal) guidelines on top. The guiding principles *should* be timeless, going forward. The need-specific guidelines (this will make more sense when we're looking at examples throughout the book) may become outdated as new use cases emerge.

If all goes to plan, this book will aim to help you use AI for fun and for work while retaining the benefits, profits, creations, outputs, and efficiencies you can realize using AI. Ideally, and probably more importantly, I am aiming to provide you with a framework for seeing all the benefits while avoiding or minimizing ethical, legal, and financial risks.

Who This Book Is For

I'm assuming that if you are reading this, you are either (1) not a lawyer or (2) a lawyer who is new to, and interested in, AI.

This book is focused on a situational approach, that is, focusing on specific use cases. So, we're going to look at a host of professional and non-professional settings where AI might be useful. The idea is to guide you to maximize your benefit if you are professionally in those fields or using AI non-professionally, while avoiding the landmines which might otherwise make you wish you had done things *the old fashioned way*. The world is always moving on and forward from whatever the *old fashioned way* might be in each scenario, but it can be hard to know what the best ways to move forward actually are which would make your life easier or more efficient.

Key professions we'll examine include

- Academia – both for students and teachers at all levels

- Sales and marketing

- The creatives – writers, musicians, and other artistic fields

- The professions – services we all use such as law, accounting, and even medicine

What should become apparent, whether you are reading this book as a whole or focusing on chapters which affect you directly, is that there is tremendous overlap and repetition in themes. AI safety isn't rocket science, but there are a host of people – many of which are lawyers – for whom it's in their interest to overcomplicate much of this subject. That said, in some less-common situations, nuances can surface or develop that require a departure from general principles. This is rare. This book aims at the general situations and use cases, which I would estimate are experienced or could be experienced by the vast majority of readers.

So, we will also examine at the holistic level whether humans are even capable of competing with AI in the long term. If we are, how – and what can we learn or predict is most protected/future-proof based on the frameworks of the law?

Finally, if you want to skip right to the back of this book, I've aimed to set out some helpful visual aids. After reading this book, hopefully you will find yourself making increasingly regular decisions as to how and when to interact with AI. These visual aids are designed to be a cheat sheet of sorts, so accelerate your recall of the risks and opportunities this book contemplates.

Disclaimer

This wouldn't be a book written by a lawyer if I didn't say – the contents of this book should not be relied on, or your sole guide, as to the legalities in using AI. If in doubt, always seek out the professional advice of a lawyer or legal professional in your jurisdiction. It rarely hurts to do your own research and gain a deeper understanding, but that doesn't mean there isn't real value in the professional advice of an independent third party.

Much of the time, if a legal issue springs up relating to your use, or the use by others, of AI, it will be (as we will see throughout this book) an issue with a solid footing in non-AI areas of law. This means that lots of intelligent lawyers can help you and steer you, given the right context and explanation. One of the key aims of this book is to help you understand that context and make informed decisions on what works best for you. But, the ultimate choice is always yours – even if that means avoiding AI entirely (not that avoiding AI is really possible now, and certainly not in future).

With that said and out of the way, let's dive in to why AI might be helpful for you and those around you.

CHAPTER 1

Is AI for Me?

The question of whether AI "is for me" is a false starting point. AI systems are of a nature that their integration into daily life is here, accelerating, and inevitably will touch most aspects of our lives. So, in the same way that you may prefer cash to card and think digital banking "isn't for me," you may well discover (if you have not already) that there are many circumstances where you are simply no longer able to have a choice if you want to access many products, services, and even utilities.

So, you can run, but you can't hide. The difference between the real world and a horror film, however, is that you don't need to hide – it's coming for you with as-yet-unknown possibilities and benefits, so long as you're able to maintain awareness of the key risks and navigate those effectively.

Am I Already Using, or Being Used by, AI?

Let's start with reality – whatever your job is, even if you're unemployed or self-employed, AI is touching your life in one way or another. Likely, in dozens of ways . Whether that is when you search for something on Google or Bing, when you are choosing your next holiday and a well-placed advert has been put in front of you, or possibly even how you are treated by governments, banks, and even your healthcare providers.

Before diving any deeper, it's worth pausing and considering the main types of AI which underpin the rapid technological advancements we are seeing expand around us and permeate our daily lives. While the introduction to this book touched on some of the key jargon, some of the

© Harry Borovick 2024
H. Borovick, *AI and the Law*, https://doi.org/10.1007/979-8-8688-0400-7_1

key terms are worth unpicking. It is an unfortunate reality that many of the key terms and categorizations of types of AI vary in how people consider or commonly use them (e.g., in academic research vs. artistic design), but the following terminology summaries are designed simply to help your further practical understanding of this book.

Machine Learning

Most modern AI relies on "Machine Learning" or "ML." IBM defines ML as "a branch of artificial intelligence (AI) and computer science that focuses on the using data and algorithms to enable AI to imitate the way that humans learn, gradually improving its accuracy."[1] For the sake of simplicity, all of the AI systems referred to in this book utilize some form of ML.

Increasingly complex and capable ML systems have impacted the lives of most people in the world for the better part of two decades. Technical advancements since 2017 have been astronomical, with some serious technical leaps (no need to know this, but should you wish to do further reading – "*Transformer-based ML*" and "*Generative Pre-Trained Transformers*" have been the technical game changers). For most people, this doesn't matter. What matters is that the quality of the interactions with daily technology has improved markedly based on three key types of AI systems.

Analytical AI vs. Generative AI

There are already hundreds of resources, including some phenomenal books, which do a good job of summarizing the main types of AI. However, it is often most useful for a reader who is not a computer scientist to focus on the use case and outcomes, rather than the underlying technical complexities, to form a coherent understanding of the different types of AI.

---

[1] IBM, "What is Machine Learning?", www.ibm.com/topics/machine-learning.

Analytical AI

The first type, and in many ways the foundational root of all current AI systems, is analytical AI. Arguably, the simplest AI often actually does most of the complex thought that other AI systems then expand upon.

In practice, analytical AI is most easily summarized by the following markers:

1. An underlying set of data.

    This is typically either text, images, or videos, but is theoretically any kind of data.

2. An algorithm applied to the data.

    The algorithm is capable of operating within the given data set, or multiple datasets. The information is systematically characterized and sorted into grouped data of similar kinds (i.e., a system which, when presented with lots of images of forests, is capable of categorizing similar things such as trees, plants, or animals as being distinct).

3. Typically, some correction is applied.

    Some re-enforcement of learning takes place, ideally addressed accuracies and inaccuracies of how the algorithm has categorized or understood the data. This can be carried out by another technical system in an automated way, or by humans.

4. A user interface of some kind which can deliver
    useful or specifically sought information as
    an output.

    Some kind of human-readable user interface is
    typically used to help people understand the data or
    inferences from the data that the AI system is able to
    identify or illuminate. This could be in the form of a
    report, graphs, or other kind of summary.

Some types of analytical AI are also referred to as "research" AI, when applied to larger or less obviously defined datasets. Terminology remains somewhat inconsistent internationally, as well as when used by technical specialists rather than ordinary end users. For example, when looking at AI in the case of professional services such as law, "analytical AI" will often be used in connection with a closed and specific set of data such as a lawyer collating large amounts of contracts to understand their contents. By contrast, "research" AI is likely to refer to the search and resulting output from a wider or less-qualified dataset, such as a broad search to find information from a typical search engine that finds a relevant website or article. By contrast, in academia, it is common, though confusing, for analytical AI (among other types of AI) to be one type of research AI[2] (i.e., where the use of "research" is intended to mean any type of AI that can be applied for academic research purposes).

Generative AI

Generative AI is the type of AI with which most people will be familiar, or will have knowingly interacted. The most famous types of generative AI are responsive chatbots such as ChatGPT or non-text content creation

---

[2] Sarke, I.H. National Library for Medicine, National Centre for Biotechnology Information www.ncbi.nlm.nih.gov/pmc/articles/PMC8830986/.

tools such as DALL-E or Midjourney.[3] The basis on which "Gen AI" works is fundamentally the same as analytical/research AI. However, it is distinct in that **"new" information is generated for the user's consideration as a result**. In reality, this works by using one or more given datasets to make a prediction as to what a specific output should be or look like. For example, predicting what the appropriate next word in a sentence should be (expandable to very large volumes of text), or predicting what a desired image should look like and generating it on the basis of a text prompt.

The real-world presence of AI, more than we realize

With a clearer understanding of the broad types of AI in mind, it is easier to start thinking about how AI may be touching our daily lives, even when using digital tools or services which we do not immediately or inherently identify to be AI-driven. Notably, the "actions" we perceive technology to be carrying out have advanced at such a pace that they increasingly cause suspicion and confusion for real human users. The recommendations we receive from technology (in various forms such as social media, online shopping, etc.) and the interactions we have with technology via chatbots and other related technology can be convincing and often we may believe we have arrived at organic or optimal conclusions, when in reality we may have been directed to specific or engineered outcomes.

AI systems which can "deepfake" to create false or misleading images , or chatbots which speak to us in a way which is a bit too real, often run the risk of either unnerving us or being convincing to a potentially concerning extent (i.e., we may be led to believe there is just one answer or one truth).[4] Many of us have also experienced that sensation of confusion or suspicion that we are being recorded when we speak about a subject and

---

[3] Mascellino, A. Technopedia, www.techopedia.com/definition/midjourney.
[4] Rozsa, M. Salon.com, www.salon.com/2023/04/15/deepfake-videos-are-so-convincing--and-so-easy-to-make--that-they-pose-a-political/.

related adverts pop up on our phones. It is easier for our minds to believe that we are being closely spied on in a very particular way, rather than the more simplistic and logical reality that we are generally predictable, and AI systems are increasingly good at reducing the sum of our potential interests to a formula based on what we do, where we live, who we interact with, etc., based on the data that we all consent to give in exchange for our digital services.[5] While such a level of data collection may still be argued to be "spying," it is more challenging to coherently argue against a system where our desires as users/consumers are calculated on the basis of the data we (often freely) give away, rather than more intensive or underhand surveillance. This creates a potential "legitimate interest" (a legal term with specific meaning and impact) for the use of our data and the provision of more targeted services using the information that we have shared. Within the framework of this book, this makes understanding our own legal rights and the impact on the services we enjoy difficult to reconcile.

For example, if the auto-correct features on our phones did not learn from our interactions, we would consider them stupid and frustrating. However, because they use algorithms (in some cases, quite sophisticated AI), we broadly accept that they are helpful and therefore do not object to their improvement – which relies on quite literally examining every word we type.[6] We often presume that the lessons autocorrect takes from us (i.e., when we ignore it, or accept it) are local and specific to us, but this may not always be the case. And yet, it is unlikely that any reader of this book will recall explicitly giving consent for the creators of any autocorrection software. Whether that is for them to be able to read/examine and utilize our words for the purposes of improving their tools, or even simply for the

---

[5] Jones, J. Scientific American, www.scientificamerican.com/article/ai-doesnt-threaten-humanity-its-owners-do/.

[6] Tableau.com, www.tableau.com/data-insights/ai/examples#:~:text=set%20fare%20rates.-,Text%20editing%20and%20autocorrect,and%20other%20text%20editing%20software.

purposes of our daily usage. That is likely because the relevant provider believes they can rely on the basis that there is a mutual legitimate interest (a useful legal mechanism), rather than requiring explicit user consent.

Technology, AI-related or AI-based tools being no exception, exist to benefit a human actor. Whether that actor is you, or you believe it to be you, is often less clear. Even a recommended advert which you loved and went on to buy the product, served at least two other parties – the advertising host (e.g., a social media platform like Instagram), and the shop from which you made your purchase. Sometimes, figuring out the real beneficiaries of an interaction or use of AI can feel opaque. For example, an image or video which is deepfaked (visually manipulated) is typically designed to misinform, which makes identifying beneficiaries and creators potentially very challenging. However, as we'll discuss throughout this book, that lack of clarity is not always an inherently bad thing. If we're smart, we can harness it for our individual benefits, in many circumstances. There are also tools that we can use to help us identify when AI has been used,[7] allowing us to better protect ourselves commercially and legally. Nevertheless, we all need to make peace that the ship has left the harbor – and our lives are exponentially being affected by AI (or at least algorithms of some kind) which we can't always identify or fully understand.

It is also worth remembering that as the nascent technology evolves, a real uncertainty arises around the ethics of our interactions and the effect on us by AI systems. So, I would strongly encourage you to consider each opportunity or risk you may be approaching where AI is a potential solution in a very particular way. Ask yourself, with respect to the use of AI to solve your issue or accelerate what you do, firstly – "Should I?" instead of simply "I can, so I will."

---

[7] OpenAI, https://openai.com/index/understanding-the-source-of-what-we-see-and-hear-online.

We will return to the matter of ethics at various points throughout this book. However, to set the tone, we're in for some big (ranging from psychological to environmental[8]) problems as a society governed by the rule of law if vast corporates, governments, and other important bodies (e.g., regulators) which are using AI, or will come to use AI, don't stop and ask themselves the same questions. The way that practically happens, is by the individuals within those organizations doing so, and that becoming a positive, perpetual, mass of ethical and societal consideration and re-consideration. This can certainly be achieved in a constructive way, which will vary from organization to organization, region to region, without obstructing business growth and innovation. It is equally important for users of AI to hold ourselves to the same ethical standards and those we expect of, and hope for, from Big Tech.[9]

Where Do We Start to Identify the Risks of AI?

If AI were a person, Machine Learning would be that person's brain. The types of Machine Learning (for example, transformer-based models or simple "old-school" algorithms) are the inner workings of that brain. Of course, this is an over-simplification,[10] but it is a useful frame to understanding. With people – we know there is a person and a brain, but

---

[8] Titcomb, J. The Telegraph, www.telegraph.co.uk/business/2024/05/05/ai-boom-nuclear-power-electricity-demand/.

[9] Moss, E. and Metcalf, J. Harvard Business Review, https://hbr.org/2019/11/the-ethical-dilemma-at-the-heart-of-big-tech-companies.

[10] de Boer, M. Human Brain Project, www.humanbrainproject.eu/en/follow-hbp/news/2023/06/29/new-algorithms-enable-artificial-intelligence-learn-human-brain/.

we don't know, or have visibility of, how their brain works. As a result, it is hard for us to know what a person is good or bad at until we engage with them. In most cases, it's very hard to ever actually understand those inner workings. This presents risks. We are less likely to go to a stranger for an important task or matter, than someone we know who has built a cache of trust with us or someone else that we already trust (i.e., a friend of a friend). Some of that trust, as well the ability to provide useful interactions with us, is depending on the context in which that person is existing.

Figure 1-1 is my own riff on how the OECD explains and illustrates how context affects inputs and outputs for people and for AI systems.

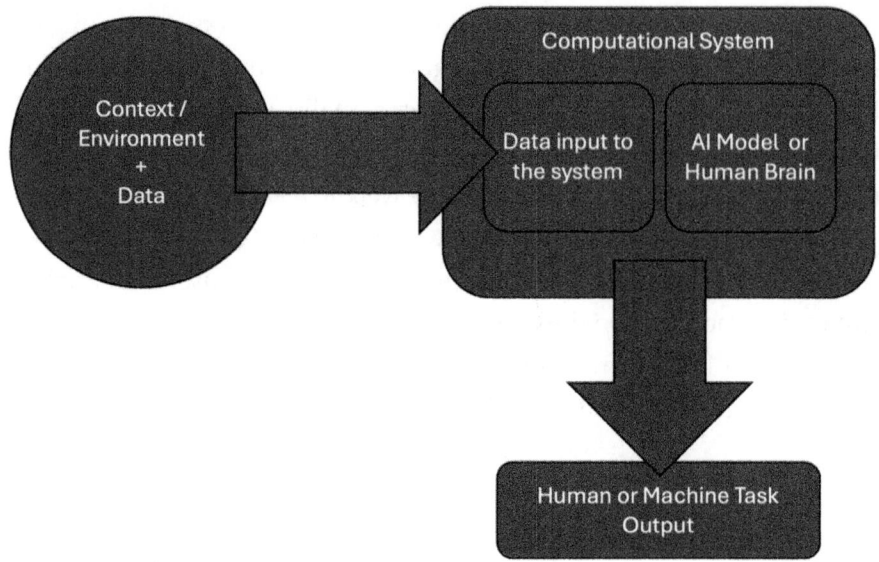

AI system

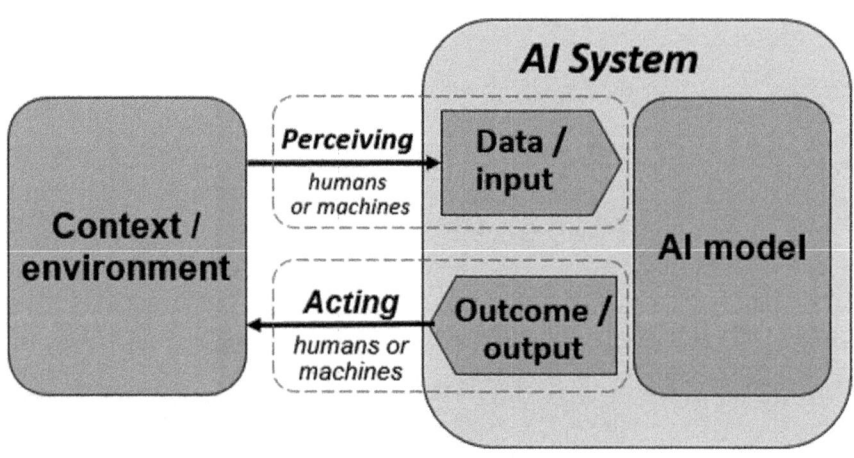

Figure 1-1. *Context and inputs/outputs*

So, AI systems are fundamentally similar to, and should be thought of in a similar way to, the ways in which we evaluate our levels of trust in other humans. Trust in AI systems, and our ability to critically evaluate the risks from that, should be based on the following parameters:

The Five Key Questions

1. **Origin**

   a. Who built the system?

   b. Human equivalent: Can we trust they are who they say they are, and do we understand their background, for example, cultural differences to understand and adapt to/for?

2. **Operation**

   a. Who runs the system?

   b. Human equivalent: Do we trust they work for, or are acting on behalf of, themselves or for who they say? Or, do they have other agendas beyond the obvious or what is presented?

3. **Access**

   a. Who has access to the system?

   b. Human equivalent: Do we know if anyone else is listening to the conversation or watching the interaction? Or, can we trust that they will/won't repeat what we discuss with others within the scope of our expectations?

4.   **Benefit**

a.   Who makes money from the system, and/or who can *conceivably* make money from the system – even via usage rather than ownership or control?

b.   Human equivalent: Do we know if someone else is incentivized or capable of taking advantage of this conversation or interaction?

5.   **Trust**

a.   Do we trust these people (i.e., the system builders and operators)?

b.   Human equivalent: Do we trust these people holistically?

Throughout this book, we will discuss industry-specific or use-specific risks and opportunities with AI systems, but they all root back to these overarching questions. In real life, we could be participating in risky activities, but if we do so with a trusted friend, then the practical likelihood of it coming back to haunt us may be lower. Or, at least we would hope.

At the very least, these foundational questions as to whether AI systems may be risky for you can allow you to know *who* interacts on the other side of your risks, even if you don't fully understand *what* the risks are.

A clear and simple example of applying these questions is presented here (ChatGPT public-facing free version as at December 2023**):**

1.   Who built the system?

OpenAI, which is owned by a series of investors, including Microsoft.

2. Who runs the system?

    OpenAI, with the support or inputs of thousands of stakeholders, including investors. Arguably, every person who gives the public-facing version of the system data is contributing to the "running" of the system. So, potentially billions of people under some form of control/operation of OpenAI.

3. Who has access to the system?

    The public-facing version of ChatGPT is fully open to the public, with more limited or advanced versions of the system being available to commercial users (i.e., there is a system which is at least two-tier). The private/admin level access is reserved for OpenAI, and potentially anyone else to whom it determines it can legally/reasonably give access.

4. Who makes money from the system?

    OpenAI, its investors, and anyone who can use the inputs and outputs of ChatGPT (including what you put into the public-facing version) for their own financial or personal gain. While we don't know for certain, we can understand from OpenAI's website, and documents imply that they do not actually onsell the raw data which is input by users to third parties.[11]

---

[11] https://help.openai.com/en/articles/7039943-data-usage-for-consumer-services-faq.

5.   Do we trust these people?

The answer will vary for each of us based on our risk appetite and trust in "big-tech." However, even for those who may believe that OpenAI has altruistic aims of advancing humanity through AI/ChatGPT, it is still a company which aims to survive and therefore make revenue. Whether or not being actually profitable is a goal is unclear. Accordingly, the very foundation of our trust should be the same as assessing trust in many other business interactions, that is, how am I making them money, in the long-term, and do I care/will it hurt me? In this case, as a free user member of the public, I am benefitting OpenAI when using their system by contributing my own prompts and data to their larger dataset to improve their model performance. They are transparent about this. So, it may be reasonable to place some trust in the use of their system and simply be careful – that is, moderate use for non-sensitive matters.

We'll return to more granular discussion of ChatGPT and other AI platforms when applied to specific use cases. But, this example is clearly illustrative. Hopefully, you can see that a framework for assessing risk of the provider is/should be established before getting into consideration of any specific scenarios or uses.

Whatever My Job Is, Can I Benefit from AI?

The short answer is yes. The longer answer is also yes, but with a *pause* before the yes. The reason for the pause is that whatever your role, whatever your organization, your first questions should either be – What do I/we need to achieve? Or, what do I/we want to avoid?

AI systems are tools. We're a long way from a fully autonomous AI, that is, a system with its own wants and needs, sometimes called "AGI". So, tools are accelerators rather than replacements for people. Tool selection is as important for AI systems as it is when I attempt, mostly unsuccessfully, any form of home improvements. The first question should not be "Which hammer?" It should be "How do I put up those shelves, and is the hammer, drill or something else the right tool to achieve that goal?"

The dissimilarity is that tool selection, particularly for simple and familiar tasks is almost instinctive for many of us. With AI systems, or even some far more fundamental and known technologies, our selection is often far poorer due to insufficient experience or intuitive differentiation between the tools. Often, we can't even decide which method of communication is best for us or the recipient, for example, Slack, SMS, E-mail, Zoom, Teams... the list is endless. However, we do (over time) build our own preferences of whether we prefer to send a WhatsApp or an IMessage for a variety of reasons which vary based on the user and the scenario.

This demonstrates two things:

1. Decision paralysis comes from an excess of choice combined with a lack of rationale for appropriate selection.

2. Knowing the right tool for the job is more likely when we are more familiar with the tools which are available and can therefore identify and utilize/avoid their strengths and weaknesses.

So, knowing the extent to which you can benefit from AI at all, or whether a particular AI system is the appropriate tool to help you for a given task, needs to be examined on a situation-by-situation basis.[12] But, there is an increasing volume of circumstances where the answer to the important question of "Is an AI system the right tool for the job?" will be yes. This will only increase over time as the capabilities of AI systems increase, and the likely financial cost to the user decreases.

The Personal Value

It's easy to over-focus on professional applications of AI and forget the daily efficiencies (or even daily joys) that AI can support in our personal lives. Just some basic conscious (i.e., non-automatic/background) applications of AI for day-to-day tasks which are already easily, and in most cases freely, available:

1. Organizing a to-do or checklist list (no, AI didn't help with this one).

2. Helping to write or co-write a speech.

3. Suggesting locations for meals.

4. Advice for things to do when visiting a new country or city.

5. Brainstorming names for a dog/baby/project.

6. Helping to write jokes or stories.

7. Support or guidance on how to do household or DIY tasks.

---

[12] UK Government, www.gov.uk/guidance/assessing-if-artificial-intelligence-is-the-right-solution.

8. Planning a workout regime or diet.

9. Research on which product to buy for a particular task/need.

10. Learning about any personally interesting subject and making that practical, for example, rather than just researching workout plans, asking AI systems to help create one for you.

11. Translation of almost any language into almost any other language (eventually, maybe even as well as a human could do when accounting for context and cultural nuances).

12. Finding new media, for example, suggesting new films or music based on what you already like. This is already often integrated into how many of our media interactions already work.

13. Simply trying to think of angles to a specific problem/solution/matter that you might not have thought about or considered.

14. Helping to weigh the pros and cons of almost any decision.

We'll explore in later chapters how some of these have limited risks when the outcomes are published, for example, if you're a professional comedian using an AI to help write jokes, then commercialize the jokes – you may need to be very careful that you're not accidentally copying someone else's material. However, for the most part, the practical day-to-day uses that many people would have of AI systems, or the ways that AI systems provide usefulness to people (whether they realize an AI is involved or not) are often low-risk. Therefore, AI systems can provide immediate positive benefits in our daily personal lives with a comparatively low chance of harm.

Summary

AI permeates our lives already, whether we are actively using it or it is operating in the background of technologies we already know, use, and accept. What matters is practical understanding and real-world mitigation of our personal and professional risks (e.g., to our privacy or confidentiality).

Hopefully, all of the discussion in this chapter is somewhat useful as a starting point for the use of AI, and the "five key questions" set out are all easily applicable (recapped here):

1.   Who built the system?

2.   Who runs the system?

3.   Who has access to the system?

4.   Who makes money from the system?

5.   Do we trust these people?

However, thinking even more practically, we may want to contextualize this to a situation and add three more questions.

If I was receiving this answer, output, help, or suggestion directly from a human being, or even a human being assisted by technology

- *Would I verify it (ask someone else/double check another source)?*

- *How much reliance (trust/faith) would I place on it?*

- *How much (money/time/effort) would I risk by being reliant on it?*

The less confident you are in your assessment and understanding of these three points, the more likely it is that then asking the five key questions with respect to the AI system is a prudent approach.

CHAPTER 2

AI and Academia

Despite much social and technological progression typically originating in academia, the education sector is undoubtedly the most unbalanced when it comes to user vs. recipient of AI-derived output. Almost universally, students move faster than educators.

Younger students move faster and typically more recklessly than more mature students. This means practically that the further away in age that students are from their teachers, the greater the gulf in harmony of perspective on when/how AI should be used as part of the academic journey. Accordingly, this almost inevitably results in an adversarial dynamic – an arms race of sorts.

In this chapter, will review whether it is actually desirable for AI to be used to better educate students and/or optimize the delivery of education. By asking this, and accepting the presence of AI in education, we can try to best determine *how* AI should factor in education in order to promote the most societally or personally beneficial outcomes. This chapter is by far the least "legal" in this book, but frames many of the practical considerations in later chapters through the lens of practicality. Legal frameworks, such as legislation or government policies, are the ultimate mechanisms by which changes to education can be brought about and enforced. In reality though, change often simply happens because of more nuanced changes in government rhetoric combined with proactive drive by educators.

© Harry Borovick 2024
H. Borovick, *AI and the Law*, https://doi.org/10.1007/979-8-8688-0400-7_2

It's important to note that when referring to "teachers" throughout this chapter, this is specifically with respect to those people actually delivering content to students, that is, those "at the coalface" of education. However, the education that students receive from teachers is impacted by a wide variety of stakeholders. Accordingly, "educators" is used to cover the broader set of people who work to create, deliver, examine, fund, and set policy for educational delivery. Separately, "legislators" are those who are often one step further removed from educators, but create the hard rules or laws within which educators have to operate and teachers must work.

Do We Want AI in Education at All?

Beware of the assumption that the way you work is the best way simply because it's the way you've done it before.

—Rick Rubin, The Creative Act[1]

The ship has sailed and the way in which students work has already experienced a fundamental shift. AI is being heavily used in academia, albeit far more on the side of the students rather than the educators. In early 2024, a survey of 1000 UK university students indicated that over half of UK undergraduates freely admit to the use of AI to assist in the creation, support, research, drafting, or revision of their essays.[2] When accounting for how truthful/conservative students may be at answering such a survey, and that the scope of the survey was limited to essay delivery only, the percentage of UK students likely to be using AI tools to accelerate their

[1] Rubin, R. (2023). The Creative Act. Penguin.
[2] Adams, R. (2024). More than half of UK undergraduates say they use AI to help with essays. The Guardian. [online] 1 Feb. Available at: www.theguardian.com/technology/2024/feb/01/more-than-half-uk-undergraduates-ai-essays-artificial-intelligence.

research and daily studies or prepare for exams is quite probably even higher. The free-to-use (in exchange for input data) nature of many AI tools is extremely enticing, but also means that this trend of rapid AI adoption is almost certainly global and not as limited by geographical wealth disparities as many other study aids or technologies.[3]

Educators at the pre-university level (broadly, kindergarten through to the end of secondary education) focus on teaching students the fundamental knowledge and concepts which frame their ability to interpret information. The desired outcome is the ability for students to then be capable of applying that information practically (and ideally, usefully) as skills are developed in later education or post-education life. A hesitation which is often commonly expressed among educators with numerous new technological developments, is that a focus on the new tool or process may bypass something which is currently considered to be fundamental. For example, whether teaching children to use word processors automatically comes at the cost of handwriting skills or if a focus on using calculators reduces basic mental arithmetic skills. However, often what teachers (and the educators who frame curriculum) may believe is fundamental is not current, but past. Teaching students how to approach problem solving on the basis of historically valued skills is actually setting them up to almost automatically be outdated by the time they enter the "real world" of adulthood and their careers. While that does not mean that some of these core skills lack practical use (e.g., long-form division or multiplication using a pen and paper), educational time is finite and must have consideration as to long-term resilience to change.

---

[3] Coffey, L. (n.d.). *U.S. Lags in AI Use Among Students, Surveys Find.* [online] Inside Higher Ed. Available at: www.insidehighered.com/news/tech-innovation/artificial-intelligence/2023/11/21/us-students-among-lowest-world-ai-usage#:~:text=While%2040%20percent%20of%20students.

With this in mind, educators and the legislators who fund and steer curriculum should be asking the following questions of knowledge and skills taught in schools and universities:

1. As best as is reasonably predictable, will what is taught be perpetually considered fundamental?

2. Can we preserve the skills and this knowledge that may no longer be fundamental for all students, for those who may still need or want them in later life?

The answer to the first question is fraught with uncertainty, and holds the potential for significant biases. A Chinese-born student who was taught to efficiently multiply numbers using an abacus[4] may have a distinct perspective from a European student taught alternative systems as to what is a fundamental mathematical skill or basic knowledge of arithmetic. It is easy to overlook these geographical and cultural differences. So, instead, it makes sense to focus on outcomes. Focusing on a particular method to secure an answer such as the correct output of an mathematical equation, may well limit the ability of students to work internationally as adults. Rather, a focus on the ability to correctly understand if the method used (e.g., a calculator) has arrived at the right conclusion, and the ability to verify that conclusion, is likely far more important. In the case of math, the desired outcome is likely to be for students to be taught sufficient mental and written arithmetic in order to allow them to (a) do basic sums using a calculator, while (b) not limiting their ability to progress to more advanced math skills.

---

[4] Anon, (2016). *Why Chinese still learn Abacus – an ancient calculation method* «*ABS Abacus Brain Study.* [online] Available at: http://absabacusbrainstudy. com/index.php/2016/01/16/why-chinese-still-learn-abacus-an-ancient- calculation-method/ [Accessed 20 May 2024].

The second question – preservation vs. progression – is surprisingly simple. Education is already layered in many countries. For example, in primary school – "basic" literacy and numeracy skills are taught. As children progress through education, it is common that based on aptitude and interest they are divided (or directed) to either advance on these subjects (e.g., advanced-level math), take alternative subjects (e.g., a greater focus on history, geography, arts, drama, or music), or combine these to some extent. Technological literacy, whether focused on AI or not, can, and should, be a fundamental school subject. Some countries, most notably South Korea, already considering technological literacy essential – with coding a mandatory school subject since 2018.[5] Of course, as AI permeates how people use technology (including coding[6]), it seems inevitable that AI naturally forms part of the same curriculum.

AI for Teachers

It is important to recognize that "learning," even within traditional school and university systems, often does (and should) have the objective of supporting students to be capable individuals in their personal lives. It is already a clear certainty that the use of AI is revolutionizing how existing careers are performed, or whether some will even continue to exist. In early 2024, the White House released a guidance note for employers

---

[5] Britishcouncil.org. (2024). *Computing (Coding) becomes required subject from next year in Korea | British Council.* [online] Available at: https://opportunities-insight.britishcouncil.org/news/market-news/computing-coding-becomes-required-subject-next-year-korea [Accessed 20 May 2024].

[6] Ranger, S. (2024). *AI coding assistants might speed up software development, but are they actually helping produce better code?* [online] ITPro. Available at: www.itpro.com/software/ai-coding-assistants-might-speed-up-software-development-but-are-they-actually-helping-produce-better-code [Accessed 20 May 2024].

and employees to understand the guiding principles on which US governmental policy will be formed around AI in the workplace. One of these principles was

> *Using AI to Enable Workers: AI systems should assist, complement, and enable workers, and improve job quality.*[7]

While this is a very soft and high-level principle, it is indicative of a recognition at the highest level of government that further legislation, policy, or at least guidelines, may be necessary for workplaces to effectively facilitate people to work productively. What this does not mention is that this should start at the root – education prior to entering the workforce, and continuing professional education. Promisingly, the UK government has led the way in piloting a "Flexible AI Upskilling Fund"; however, this is only available to adults already in full time work at small to medium sized enterprises[8] and makes no mention of any support or change for young adults or children in university or school-level education.

As a whole, statements and guidance by governments across the world regularly seem to lack acknowledgment that AI increasingly permeates the lives of regular people outside of the workplace as well. Education solely applicable in the professional spaces in which students will later engage is not particularly useful if a more holistic education is not received. For example, the abilities to communicate effectively (including strong literacy), understanding important historical events, having a rough idea of geography, learning languages – these are all skills which are as useful,

---

[7] House, T.W. (2024). *FACT SHEET: Biden-Harris Administration Unveils Critical Steps to Protect Workers from Risks of Artificial Intelligence.* [online] The White House. Available at: www.whitehouse.gov/briefing-room/statements-releases/2024/05/16/fact-sheet-biden-harris-administration-unveils-critical-steps-to-protect-workers-from-risks-of-artificial-intelligence/.

[8] GOV.UK. (2024). *Flexible AI Upskilling Fund pilot: open for applications.* [online] Available at: www.gov.uk/government/publications/flexible-ai-upskilling-fund.

perhaps even more useful, in personal lives than in many professional situations.[9] That is not to say that math or sciences are less important, but perhaps less practically necessary for day-to-day adult lives (particular when taught at their more advanced levels) when available technologies can take much of the burden.

Broadly, teachers and educators at every level of academic institutions should be asking

1. Is it desirable to use AI to better educate students and/or optimize the delivery of education?

2. Is it preferential in the medium to long term that student usage of AI is recognized and receives head-on engagement by teachers and examiners?

3. Are there long-term detrimental effects to the quality of teaching or student understanding which may arise either directly or indirectly from the use, overuse, or reliance on AI either by students or teachers?

4. Do ethical considerations arise or directly conflict with optimal or even minimal use of AI by either students, teachers, or both? This should also include consideration of third party rights.

5. Are careers for students, post-education, likely to be reliant on AI tools or assistance? Or, potentially be replaced entirely by AI?

---

[9] www.mabletherapy.com. (n.d.). *Could poor communication skills be holding students back in maths?* [online] Available at: www.mabletherapy.com/blog/2020/11/25/could-poor-communication-skills-be-holding-students-back-in-maths.

If the answer to any/all of these questions is yes – then AI training and skills, or at least recognition and consideration of the impact that AI has on current curriculum, is essential. Legislative reform can achieve this. A disappointing aspect of the EU AI Act 2024 was that, as the world's first cross border (and relatively comprehensive) AI-focused legislation, it was effectively silent on education to mitigate potential harms and maximize potential gains of AI development and usage.

Nevertheless, for educators, it is reasonable to assume that in the medium-to-long term there will be increasing governmental policy and regulation to encourage the roll-out of AI education. Where possible, educators (most importantly teachers, when given the freedom to do so) should incorporate recognition of AI in the classroom to prepare students before it becomes mandated. At the very least, educators should recognize that there is no specific legal barrier for the use of AI by students to gain advantages or accelerate their work beyond existing broad anti-cheating rules.

If educators do not recognize this and incorporate some guidance to students, a major issue will inevitably arise when students do not consider support by AI systems of their academic output cheating, in the absence of specific legal or regulatory frameworks. Quite simply – it makes sense for teachers and their students to be on the same page when it comes to what is or is not acceptable, while encouraging future-proofing in student methodologies. The law, unfortunately and currently irrespective of jurisdiction, doesn't currently facilitate this open collaboration.

Student Usage of AI

There is a recurring theme throughout this book: outcomes are generally more important than processes unless the process has some kind of emotional consideration, or unless the short-term benefit is likely to be totally negated by a long-term effect. Outcome-focused approaches are key to our professional lives, and should form the basis of the education children receive.

Conrad Hughes, Director General of the International School of Geneva, succinctly set out in September 2023 (published by the World Economic Forum):

> *Many have warned that generative AI could harm students' learning, providing them too easy a way to complete work, get answers or write. But we live in a world where architects, designers and engineers use the best technology available to them every day. We must examine whether it really is sensible to prevent students from engaging in the same behaviour in their own world.*[10]

For most, knowing the answer to a mathematical problem (no matter how potentially important) is unlikely to have a significant emotional consideration. Therefore, if the goal is preparing students for the short-to-medium term of their post-secondary education life, it is most desirable for teachers to preserve the skills to verify the outputs of digital answers (including from AI systems), without detriment to the educational progress of actually supporting students to swiftly and accurately reach correct answers and accurate understanding of the specific subject matter.

Where this becomes more complex is where there are subjectivities. For example, in the teaching of subjects such as history or religious education. It would be a poor educational outcome for a student to use an AI system (or even simply search on Wikipedia) to learn about a historical event or religious belief without an understanding that there may be significantly diverging opinions on evidence and interpretation. This is a compounding difficulty for those creating a curriculum or delivering

---

[10] Hughes, C. (2023). *Generative AI won't kill education unless we allow it to.* [online] World Economic Forum. Available at: www.weforum.org/agenda/2023/09/navigating-the-rise-of-generative-artificial-intelligence-and-its-implications-for-education/.

classes, as it may appear to require different policies and approaches
vs. the teaching of mathematics. Although there are plenty of diverging
theories in math and science, the fundamentals are broadly taught as
constants while receiving infrequent but periodic updates. This is similar
to how history is taught in many schools, that is, "this is how our books tell
us this happened." For example, a British and Russian school are likely to
have very different focuses and perspectives on the events of World War
Two, even if they were somewhat allied against the same enemy and most
of the facts are agreed.

Therefore, the key skill is critical analysis rather than acceptance of
answers. Verification of sources, and a taught understanding that sources
can be biased, and can be cross-checked, is becoming an increasingly
essential skill. A comparable example in daily adult life is that most of
us would generally accept it is not normal or sensible to drive into a
lake simply because a map on our phone tells us that is the route to take
(though that doesn't stop some people from occasionally doing so...[11]).

Where the law and education intersect here is typically in the absence
of legal frameworks. Teachers should recognize (and ideally teach their
students) that even their source materials may have biases and that
curriculum created by educators (as directed by government policy and
legislation) does not for the most part protect against or prevent this.
Legislators and government bodies have it within their gift to create legal
frameworks to support student understanding of critical analysis and

---

[11] Bharade, A. (n.d.). *Tourists in Hawaii followed their GPS and drove their car
straight into a harbor: "Pretty sure that was not supposed to happen."* [online]
Business Insider. Available at: www.businessinsider.com/tourists-hawaii-gps-
drove-car-into-water-2023-5#:~:text=A%20pair%20of%20tourists%20drove
[Accessed 20 May 2024].

bias detection. Some countries, notably Finland, have started to include training for students to recognize biases across sources – primarily with a focus on disinformation and conflicting information in media.[12]

Fundamentally, this means that teachers must make allowances for and accept AI systems in the same way that the use of existing digital repositories and searches have become accepted norms.[13] Acceptance means practically acknowledging that information from such sources will help students reach answers, but proportionally allocating time in each class (or at least each semester) to critically assess the stated facts and the certainty/reliability of those answers. More importantly, even if the facts are broadly uncontroversial or factual, it is essential for students to develop the skills to critically interrogate the conclusions which may be presented.

Failure to encourage the development of these skills at an early stage of education will inevitably result in students being left to their own devices and conclusions with AI-stated "facts" and answers. This practically means that students may clash with the presented curriculum and leave them confused as well as unable to meet the requirements of standardized tests.

AI and the Long-Term Quality of Education

Whether students are taught skills of critical analysis or not, there is a wider question as to whether AI in education will have inherent long-term detrimental effects and whether educators (or even legislators) should seek to mitigate that potential harm. This detriment may be to the quality

---

[12] Henley, J. (2020). How Finland starts its fight against fake news in primary schools. *The Guardian.* [online] 29 Jan. Available at: www.theguardian.com/world/2020/jan/28/fact-from-fiction-finlands-new-lessons-in-combating-fake-news.

[13] Kupperstein, J. (2023). *AI Can't Replace High-quality Teaching: Using the Technology as a Tool.* [online] EdTech Digest. Available at: www.edtechdigest.com/2023/06/15/ai-cant-replace-high-quality-teaching-using-the-technology-as-a-tool/.

of teaching or student understanding. Either may arise either directly or indirectly from the use, overuse, or reliance on AI either by students or teachers. Human nature is such that we tend to favor the taking of an "easier" route.

For Teachers

Use of AI by teachers to assist in preparation, teaching, and examination would certainly be easier than manual/digitally supported manual methods. However, excessive reliance on AI is likely to have three significant long-term impacts which would be viewed by many as negatives.

1. **A decline in the human touch.**

    "...in our haste to adopt AI, we must not forget the value of the human touch. While some advanced AI systems may replicate the human decision-making process, they may never access the subjective insights and instincts accumulated through years of lived experience."[14] – International Institute for Management Development

    Much of pre-university education is tied to a strict curriculum, to which teachers add their own flair or personality. Excessive reliance on AI to draft (or, foreseeably, present) lesson plans and teaching guidance is likely to lead to greater standardization

---

[14] I by IMD. (2024). *Keeping the human touch: Why people remain essential in the age of AI - I by IMD.* [online] Available at: www.imd.org/ibyimd/artificial-intelligence/keeping-the-human-touch-why-people-remain-essential-in-the-age-of-ai/#:~:text=But%20in%20our%20haste%20to [Accessed 20 May 2024].

even beyond the centralized curriculum of each school/district/nation. This risks stamping out any creative spark in teaching, as well as the excitement of pupils to learn. Excessive focus on AI as a potential replacement for real flesh-and-blood teachers, rather than as a teaching aid or as a subject to be understood by students, risks missing a major point – teachers share more than just the curriculum they teach. Teachers share their own experience and enthusiasm for a subject. Anyone who fondly remembers a subject they studied at school or university can likely tie this to a positive, encouraging or simply enthusiastic teacher.[15] In some cases, even a teacher who provides simple assurance without much enthusiasm can have a powerful impact on the student experience. It is not possible to create legal or policy frameworks which can bypass this fundamental truth. That said, in limited circumstances – for example, in remote areas or in severely impoverished developing nations, AI can provide a valuable teacher with relatively little steer, where there may be no human capable or available to teach.[16]

---

[15] www.oecd-ilibrary.org. (n.d.). *Teacher enthusiasm | PISA 2018 Results (Volume III) : What School Life Means for Students' Lives | OECD iLibrary.* [online] Available at: www.oecd-ilibrary.org/sites/h9ed2d5d-en/index.html?itemId=/content/component/b9ed2d5d-en. and Lazarides, R., H. Gaspard and A. Dicke (2019), "Dynamics of classroom motivation: Teacher enthusiasm and the development of math interest and teacher support," *Learning and Instruction*, Vol. 160, pp. 126–137, https://doi.org/10.1016/J.LEARNINSTRUC.2018.01.012.

[16] virat, virat (2023). *Transforming Rural Education: Unleashing the Potential of AI.* [online] Medium. Available at: https://medium.com/@viratviratvv2/transforming-rural-education-unleashing-the-potential-of-ai-8fd4aea78d6f.

2. **An increase in "forgotten" knowledge leading to a monopoly opinion.**

Two teachers in the same school teaching the same subject may have their own takes and knowledge (which may be outside of, but connected to, the core curriculum). The result is that students from two different classes are likely to have learned the core facts needed for their examinations, while having a divergence of experience and opinion which can lead to essential life skills. For example, the ability to understand diverse view points and have a healthy debate relating to the nuances around facts. Excessive standardization through lesson planning, if entirely delegated to an AI system, is likely to lead to less critical thinking (as already discussed, a major skill that teachers should seek to develop in students) as well as a more uniform view of facts. Whereas when entering higher education and the professional workforce, critical analysis and the ability to interrogate subtleties around the understanding of a fact are both essential skills. However, there is an important nuance. There are clear legal frameworks and duties in many countries (of which teachers should be mindful) whereby teachers and educators generally are expected to act and deliver education to students in a manner that is politically neutral.[17] In effect, there can be professional and personal sanctions, such as the

---

[17] Department for Education (2022). *Political impartiality in schools.* [online] GOV.UK. Available at: www.gov.uk/government/publications/ political-impartiality-in-schools/political-impartiality-in-schools.

termination of employment, should students receive education which is overtly politically biased. This becomes particularly complex when religious or faith-based schools overlap with political positions (e.g., faith schools and conservative political values) in secular countries.

However, the introduction of AI is actually likely to give students access to a greater diversity of sources and materials than might be found in educational institutions where teachers have relatively homogenous perspectives. In any event, if AI systems used in schooling are providing excessively politically biased information to students, the relevant educators may well be legally responsible and face consequences. Whether there are any consequences, and their severity, will vary between countries. In some countries, particularly those which are not secular, homogeneity of teaching along religious or political lines may be actively encouraged,[18] irrespective of whether this produces greater long-term educational outcomes.

3. **Potential increases of errors and omissions being missed.**

If teachers are excessively reliant on AI systems or technology generally in lesson planning and assessment, they are less likely to notice, critically assess, and remedy any failure, error, or omission

---

[18] Yusuf, M. (2020). Why Indonesia Prefers A Mono-Religious Education Model? A Durkhemian Perspective. *Al-Albab*, 9(1), pp. 37–54. doi:https://doi.org/10.24260/alalbab.v9i1.1555.

in that lesson plan. Of course, the risk goes beyond just lesson plans and the same could be said for teachers who may already place excessive reliance on template or cookie-cutter lesson plans, without appropriate review and adaptation where necessary. Such templates will often be consumed and form part of the underlying understanding of an AI system which will produce a lesson plan for a teacher. Whilst technology for accelerating the planning and delivery of classes have been around for some time, including AI systems,[19] teachers remain notably (and encouragingly) conservative about how much reliance can be reasonably placed on AI[20] – recognizing and preserving their own value as teachers.

Each of these three risks do not just apply to teachers when preparing and delivering lessons but also to legislators, government departments, and school boards when creating and updating the curriculum on which the lesson plans are based. It is also inevitable that students will face a somewhat similar dilemma of ease vs. excessive reliance on technology.[21] The main nuance is that while students broadly only owe themselves a

---

[19] www.magicschool.ai. (n.d.). *MagicSchool.ai - AI for teachers - lesson planning and more!* [online] Available at: www.magicschool.ai/#:~:text=Educators%20 use%20MagicSchool%20to%20help [Accessed 21 May 2024].

[20] Busby, E., The Independent. (2024). *Two in three teachers say AI is "too unreliable" to assess pupils' work – survey.* [online] Available at: www.independent. co.uk/news/uk/politics/chatgpt-department-for-education-university-of-exeter-dfe-yougov-b2515361.html.

[21] Ahmad, S.F., Han, H., Alam, M.M., Rehmat, Mohd.K., Irshad, M., Arraño-Muñoz, M. and Ariza-Montes, A. (2023). Impact of artificial intelligence on human loss in decision making, laziness and safety in education. *Humanities and Social Sciences Communications*, [online] 10(1), pp. 1–14. doi: https://doi.org/10.1057/ s41599-023-01787-8.

duty to perform in academia (which is entirely optional past a certain age, depending on the country/region), whereas teachers and educators typically owe some level of professional, ethical, and legal duty of care to their students. Accordingly, greater scrutiny should be applied to them as to whether their use of technology, including AI, goes beyond useful reliance and into detrimental over-reliance.

For Students

Use of AI by students to do or support with homework, planning for exams is already commonplace.[22] It is extremely hard for teachers to detect the involvement of AI systems in some of the most common applications (e.g., giving the outline of an essay plan). There is an abundance of technology, including AI systems, designed to detect the presence or usage of other AI systems in academic output. However, in practice, these are currently either too expensive to be widely available, or insufficiently accurate. Accordingly, teachers are skeptical of how effective such tools might be[23] – and students either share the same sentiment or are indifferent to the risks of being "caught." However, excessive reliance on AI systems by students may have three long-term detrimental consequences. As the purpose of education is to generate positive learning outcomes for students, these consequences should be considered as more severe, and certainly more immediate, than the risks for teachers.

---

[22] Heaven, W.D. (2023). *ChatGPT is going to change education, not destroy it.* [online] MIT Technology Review. Available at: www.technologyreview.com/2023/04/06/1071059/chatgpt-change-not-destroy-education-openai/.

[23] Coffey, L. (2024). *Professors Cautious of Tools to Detect AI-Generated Writing.* [online] Inside Higher Ed. Available at: www.insidehighered.com/news/tech-innovation/artificial-intelligence/2024/02/09/professors-proceed-caution-using-ai.

1. The risk of missing the point.

    Education has two tranches – (1) the knowledge
    to simply get the grade which may help a student
    progress to the next step of education (when
    education can feel like a series "box-ticking"
    exercises); and (2) the knowledge and skills to
    practically succeed in later life, including the
    application of some of the education which may
    have seemed like "box-ticking" at the time.

2. The potential to sabotage later academic, social, or
    employment opportunities.

    Whenever a student places excessive reliance on
    technology or any aids, they take a risk that they
    may be missing the point of their education (per
    preceding risk 1). However, usage of AI systems or
    any other technology can potentially go outside the
    bounds of reasonable educational support, efficient
    methodologies or even clever shortcuts. Should
    students' workflow go into deeper grey areas or
    worse – relies on outright cheating or plagiarism,
    they risk being found out/caught and punished. A
    direct "copy-paste" from ChatGPT is unlikely to be
    justifiable in any event, in the same way that it would
    not be acceptable to directly copy from Wikipedia or
    any other source.

    The reality is that, irrespective of whether the
    student cares about any moral or ethical grey areas,
    there are often legal lines that can be crossed.
    When these rules are breached, sanctions can go
    on permanent educational records which can limit

future educational or professional opportunities. In simple terms, the rules (and the institutions which set out the rules – that is, schools and universities) rarely care whether a student believed cheating to be wrong. They only care whether it was permitted or not within the rules they have set out, which are formed by higher-level governmental or regulatory guidelines (or even laws).[24] If anything, this may actually work to the advantage of students as it clarifies how far students can push the efficiencies and advantages of AI systems in their studies without finding themselves in trouble. On a practical note – students should keep a note (ideally with logs from the AI systems used) to show their reasonable use and avoid false or misplaced accusations of cheating.[25]

---

[24] Dickinson, J. (2024). *There's no point comforting ourselves over AI and cheating when we don't know what cheating is.* [online] Available at: https://wonkhe. com/wonk-corner/theres-no-point-comforting-ourselves-over-ai-and-cheating-when-we-dont-know-what-cheating-is/ [Accessed 21 May 2024].

[25] www.euroeducation.net. (n.d.). *How to Use AI Tools for Studying (Not Cheating).* [online] Available at: www.euroeducation.net/articles/how-to-use-ai-tools-for-studying.htm.

3. Boredom, and further disinterest in the subject matter as a result.

    Any time a student has the opportunity to "get to the answer" of any academic problem, or skip to the answer of any educational challenge,[26] they risk only delivering the answer without developing an interest in the actual subject. Technological shortcuts, such as the use of calculators instead of long-form division/multiplication may be time efficient, but they do not drive students to engage in a subject, nor teach a student how to solve a problem using only their mental resources. This means that students are less likely to see the "magic" in some academic subjects and want to progress to more advanced levels (e.g., from multiplication progressing to algebra and statistics). This may also limit the ability of students to participate in complementary subjects – for example, in the sciences which interdepend on math.

An example of how this could actually impact job prospects is in the hiring process for many graduate roles. The use of problem-solving questions which are purposefully hard, or impossible, to answer is commonplace. This is particularly typical at large tech firms or investment banks who may ask a seemingly trivial questions such as "How many golf balls can you fit inside a 747 Jumbo Jet?" The point of such a question is primarily to explore mental problem-solving, enthusiasm, and approach to hypothesis-generation – in some cases, almost irrespective of the

---

[26] Dickinson, J. (2024). *There's no point comforting ourselves over AI and cheating when we don't know what cheating is.* [online] Available at: https://wonkhe.com/wonk-corner/theres-no-point-comforting-ourselves-over-ai-and-cheating-when-we-dont-know-what-cheating-is/ [Accessed 21 May 2024].

accuracy of the answer. So, students should be mindful that over-reliance on AI, like over-reliance on any tool which allows them to skip to the answer, is likely to be short-term beneficial at best, actively detrimental in the long-run at worst. The solution is quite simple – to be mindful of learning the underlying skills, while also taking the fastest route to solve problems when necessary or the critical-thought skills have been built as a foundation. Practically, this may mean solving the academic problem swiftly with the support of AI systems and revisiting the matter to consolidate genuine understanding. Of course, students should be mindful that they may need to redeploy that understanding in a situation where no AI system is available, for example, a formal examination.

Ethics and AI

It is too easy to simply equate the use of AI systems by students as a form of cheating or symptomatic of laziness.[27] With respect to teachers, there is a common narrative that the use of AI systems (or tech tools generally) may be equally lazy or contrary to the ethical duty teachers have to deliver a quality education to students. However, each of these perspectives is either entirely false, or detrimentally reductive.

The legal frameworks around educational integrity, in most countries with developed and nationally standardized education systems, allow educators at the individual school/university level considerable scope to determine whether student or teacher conduct is ethically acceptable. More serious matters may be escalated or appealed to governing bodies, but most disciplinary issues are commonly dealt with in the here-and-now at a local level. This locality and immediacy benefits from current

---

[27] Samilu, B. quoting Kakumba, U. (2023). *Artificial Intelligence making our students lazy – Mak don.* [online] Monitor. Available at: www.monitor.co.ug/ uganda/news/national/artificial-intelligence-making-our-students-lazy-mak-don-4404676.

context and rapid resolution. A rationale for this setup, other than simple proportionality and appropriate use of resources, is that most disciplinary matters are either minor or possible errors of judgment.

Concerted major efforts to "beat the system" are far less common, though certainly very real,[28] so the legal and enforcement mechanisms are arranged accordingly. This means that if a student or teacher potentially breaches their ethical obligations or limitations, they should (on the whole) expect a proportionate response. This proportionality should apply irrespective of the mechanism/form, unless an academic institution is aiming to set an example of a student or teacher to prevent specific future behavior. There is a real possibility that, particularly with respect to student usage of AI systems, academic institutions may impose seemingly disproportionate punishments/sanctions with the intention of visible deterrence.

Students

The reality is that students have had to defend simplistic accusations of cheating on their homework since the early 1990s when the Internet first started containing academic libraries, more in the late 1990s when search engines first emerged, and hitting a high point of teacher/parent concern in the early 2000s when Wikipedia first became an easily accessible repository of summary information. Many millennials may remember the decision as to "just how much can I trust/copy Wikipedia for this homework?"

---

[28] Pérez-Peña, R. (2012). Studies Show More Students Cheat, Even High Achievers. *The New York Times*. [online] 8 Sep. Available at: www.nytimes.com/2012/09/08/education/studies-show-more-students-cheat-even-high-achievers.html.

However, two major shifts have occurred in the past 20 years:

1. Significant improvements in accuracy, reliability, engagement with and verification of online resources such as Wikipedia.[29] It is easy to forget that even the helpfulness of even simple resources such as Google searches have significantly improved at a steady pace.

2. Students who are born post-2000, who have only known the era of easily searchable and accessible online information. Without the pre vs. post-Wiki hesitation, students are more comfortable relying entirely/nearly entirely on searchable resources.

A 15-year-old student in 2024 (when this book was written) is unlikely to even consider the ethics of whether they should be heavily using, reliant, or even dependent on online resources rather than schoolbooks or classroom-prescribed online information.[30] The same might even be said for young teachers.

That same student will never have experienced a comparative era before easily accessible and diversely sourced online resources. So, the ethical considerations that many Millennials (or older) may have had drilled into them throughout their education are simply unknown-unknowns to Gen-Zs (or younger).

---

[29] Dewey, C. (2016). The surprising reason some college professors are telling students to use Wikipedia for class. *Washington Post.* [online] Available at: www.washingtonpost.com/news/the-intersect/wp/2016/06/20/the-surprising-reason-some-college-professors-are-telling-students-to-use-wikipedia-for-class/.

[30] Henry, J. (2022). *Students Are Using AI Programs to Finish Their Assignments — Is it Considered Ethical?* [online] Tech Times. Available at: www.techtimes.com/articles/280169/20220906/students-using-ai-programs-finish-assignments-%E2%80%94-considered-ethical.htm [Accessed 20 May 2024].

AI systems have only been easily available to the public since late 2021/early 2022. Pre-teens, teenagers, and university students are typically early tech adopters, particularly when it can help them secure better grades. Even more so, when better grades can be obtained with equivalent or less work. Already, we can see a rapid adoption and regular usage – way ahead of the ability of educators to detect usage, and seemingly too fast for teachers to even productively collaborate with students as to how best to manage their usage of AI tools.[31]

Various surveys throughout 2023[32] have shown that between 30% and 50% of university student have used ChatGPT or other systems on an occasional but regular basis, with up to 89% of school students regularly using AI systems to accelerate or support their homework. Up to 53% of students have admitted to using AI systems to write either a full or a significant part of an essay, whereas only approximately 22% of students have used AI systems to assist in an essay outline.

These numbers should be deeply concerning to teachers. They demonstrate a lack of boundary-setting and communication between teachers and students as to what is acceptable. More concerningly, it is unlikely that many teachers are remotely aware that such reliance on AI systems by students is already in effect. This creates its own ethical dilemma – a fair and well-governed educational system depends on an understanding by educators as to how students are tackling the tasks placed before them. Students broadly should already be aware that

---

[31] Lieber, R. (2024). U.S. NSF National AI Institute for Student-AI Teaming. *The Importance of Teacher Collaboration When Developing AI Partners for Education.* [online] Available at: www.colorado.edu/research/ai-institute/2024/03/18/importance-teacher-collaboration-when-developing-ai-partners-education [Accessed 23 May 2024].

[32] Westfall, C. (2023). *Educators Battle Plagiarism As 89% Of Students Admit To Using OpenAI's ChatGPT For Homework.* [online] Forbes. Available at: www.forbes.com/sites/chriswestfall/2023/01/28/educators-battle-plagiarism-as-89-of-students-admit-to-using-open-ais-chatgpt-for-homework/.

simple plagiarism is not acceptable and should be met with negative and potentially severe consequences. Many students are also technologically savvy enough to understand that AI systems draw from existing online information and datasets to provide tailored answers. So, it is not a huge leap of logic to assume that most students should understand that using AI systems to generate full essays is likely to be equivalent to plagiarism.[33] Yet, at this relatively early stage of adoption, some students appear predisposed to excessive or full reliance on AI systems (even if they make tweaks). Rather, if AI systems are to prove successful in academic environments, students need to take some responsibility and recognize that AI systems should be used as supportive tools to provide outlines, summaries, plans, and basic answers on which the students can expand. The use cases and helpful applications are likely to constantly broaden, but the ethical lines of AI systems "doing the work" vs. "helping the student work" should remain relatively consistent.

The likely long-term practical bargain is that obvious incidents of full-blown dependency on AI systems will need to be met with prohibitive consequences by educators, and students will need to engage with their teachers so as to periodically gauge what may be considered excessive usage and what may be reasonable. Where these lines of acceptability are will shift as the understanding by educators, and their tolerances, advance. By engaging transparently in this process, students gain the safety of knowing what they can do and utilizing that to its fullest extent.

Educators

What may be the most difficult for educators at every level to accept and understand is that mild-usage, or even full-blown dependency, on AI systems may not be obvious. Even beyond the experience and skill of teachers, there are also many tech tools which claim to be capable of

---

[33] www.nytimes.com/2023/02/02/learning/students-chatgpt.html.

detecting AI-derived content, with varying degrees of success. As little as 1 in 10 teachers "admit" to not being able to identify the AI-generated vs. student-generated work.[34] However, other studies demonstrate that the reality is that this number is closer to 2 in 5 teachers actually having any ability to detect AI.[35] This demonstrates that a major initial stumbling block to a bridge between educators and student collaborative use of AI is recognition of its use.

Realistically, as the sophistication of AI systems increases, which can be used for completing academic tasks, it is likely that the ability of educators (particularly examiners or teachers marking assigned work) to detect or resist AI-derived academic work will diminish. This is most probably irrespective of any advances in the "counter-AI" or "AI detection" tools.[36] So, is attempting to detect AI-generated or AI-derived academic work even worthwhile? Broadly, no – other than for the sake of a *somewhat* visible deterrent

For at least 15 years in universities, and at least ten years in many UK, US, and European schools, basic plagiarism-detection tools have already been in regular use. This anti-plagiarism verification typically happens at the point when a piece of homework, assignment, or dissertation is submitted. So, we already have tools that can detect the most simple

---

[34] Editor, F.N. (2023). *1 in 10 teachers can't tell AI and student work apart - training + regulation now essential.* [online] FE News. Available at: www.fenews.co.uk/education/1-in-10-teachers-cant-tell-ai-and-student-work-apart-training-regulation-now-essential-according-to-research-and-edtech-expert/ [Accessed 23 May 2024].

[35] admin (2023). *Study finds that three in five teachers can't identify AI written content – The Educator Magazine UK.* [online] Available at: www.the-educator.org/study-finds-that-three-in-five-teachers-cant-identify-ai-written-content/ [Accessed 23 May 2024].

[36] Coffey, L. (2024). *Professors Cautious of Tools to Detect AI-Generated Writing.* [online] Inside Higher Ed. Available at: www.insidehighered.com/news/tech-innovation/artificial-intelligence/2024/02/09/professors-proceed-caution-using-ai#:~:text=The%20Effectiveness%20of%20AI%20Detectors&text=That%20same%20month%2C%20a%20team.

and significant concern – straight-up copying. Assuming that these tools remain somewhat accurate, it is important that this remains the primary frontline concern to educators. This is not only for the benefit of students, but because of the potential ethical and legal consequences, and impacts on the rights of third parties such as writers and publishers. Their rights and concerns also need to be considered, irrespective of whether a student is doing a simple copy-paste, or when their content is fed into AI tools which generates a substantially similar output to meet a student-made prompt.

Careers and AI

A core societal role of teachers is to use academic education to prepare young people for the intellectual, and in many cases broader, challenges of adult life. Most importantly, the skills and understanding needed to have a fair shot at securing and maintaining employment.[37] This should be the absolute baseline. In practice, teachers are constrained by varied limitations – budget, time, mandated curriculum, and the willingness of students to engage with taught subject matter. AI can potentially help fill in some of the gaps – either as a skill-based subject to taught stand-alone, or as methodology in the study and comprehension of other subjects.

Typical core/fundamental school subjects such as math, languages, sciences, history are important and should remain in academic curriculum. But, it is worth remembering that societal perceptions of a subject or methodology as "fundamental" can shift, and that the education sector has historically been slower to adapt than the professional world. For example, many British schools used to (some still do) teach Latin

---

[37] Mearian, L. (2023). *The most in-demand AI skills — and how companies want to use them.* [online] Computerworld. Available at: www.computerworld.com/article/3705095/the-most-in-demand-ai-skills-and-how-companies-want-to-use-them.html.

as a mandatory subject.[38] Many people would look at this as comically backward-facing, though arguably well intentioned. Historically, lawyers and even doctors were required to know some basic Latin terms, so by dropping the subject a school would have risked actively prejudicing their students from achieving within the legal or medical professions. Those professions, like all industries to varying degrees, have had to update and adapt to the changing world around them. Latin is now nearly useless for a lawyer, and the use of Latin in legal conversation will now almost certainly lead the listener to assume that the person using a Latin term is doing so because they don't understand the fundamental meaning of what is trying to be expressed.

So, we should accept that certain subjects can be introduced to changing school curriculums over time and others can become less useful. At the very least, if the subjects remain the same, the central themes and knowledge within those subjects may be slimmed down or adapted.[39] Accordingly, the introduction of AI into the school curriculum is potentially very low-friction. It is entirely sensible for schools to consider teaching more advanced AI skills (coding, prompt engineering, advanced computer science). But, the low-hanging fruit is simple acceptance by schools that in researching questions such as "why did WW1 break out?" for a history class assignment, or "How do I understand a balance sheet?" for an economics project, schools should be permitting and encouraging students to do so in the way that they will be in their careers, however close to "cheating" schools may consider this.

---

[38] Woolcock, N. (2023). Latin is now fourth most-taught language in primary schools. *www.thetimes.co.uk*. [online] 5 Nov. Available at: www.thetimes.co.uk/article/latin-language-lessons-uk-primary-schools-2023-wrqrtfj0s [Accessed 5 Nov. 2023].

[39] Elliot, E. (2023) HMC (The Heads' Conference). *AI and the future of education*. [online] Available at: www.hmc.org.uk/blog-posts/ai-and-the-future-of-education/.

As already outlined, there is a big difference between a simple copy-paste attempt at plagiarism vs. using AI to help comprehension or set frameworks and outlines for academic tasks. The latter should be encouraged if schools and universities want to honestly prepare students for the real world. Failure to do so holistically, would simply give other students who have been more active and developed skills in using AI to improve their academic output (without straight plagiarism) a competitive advantage in the workplace. Considering that some countries' education systems are already far more tech-focused or tech-enabled, this could lead to international skill imbalances between nations who embrace AI skills and AI usage into the educational norm, against countries where this is frowned upon or considered to be unsuitable for academia.[40]

It's an over-used and overly simplistic phrase, with many variations (and for the life of me, I can't figure out who genuinely coined it), but the following is inevitable and educators must accept it:

AI may replace some jobs, but for many jobs – people who are willing and able to use AI will simply replace or surpass those who don't.

Key Considerations
For Educators

- Can we do more to teach students how to use AI tools as a study aide or accelerator for any subject without compromising their ability to understand the core points of the lesson/subject?

---

[40] Rigley, E., Bentley, C., Krook, J. and Ramchurn, S.D. (2023). Evaluating international AI skills policy: A systematic review of AI skills policy in seven countries. *Global Policy*. doi: https://doi.org/10.1111/1758-5899.13299.

- Should we incorporate AI-specific skills as part of our STEM curriculum, or as a tool/methodology within every subject?

- If we are considering the value of teaching students how to use AI tools as a study aide or incorporating AI-specific skills into the curriculum

- How can we do this in our specific circumstances in the most effective and least disruptive manner (e.g., can we use free-to-use tools to minimize cost-based obstacles)?

- Are we doing the best for our students in preparing them for how they will practically research, navigate, and solve problems in the workplace?

For Students

- Is my use of AI tools actually helping me deliver a better quality of academic work, or just do a similar/worse job faster?

- When I use AI tools, am I still understanding the core points of the lesson/subject – or am I just skipping to the answer?

- Does the way I am using AI tools risk setting me up to be over-reliant in a way which is likely to be less helpful in later life/in the workplace? Or, am I using AI tools in a way which is likely to still be acceptable in the workplace?

- Am I using AI tools in a way that is likely to simply be considered cheating/plagiarism (whether now or when I'm in the workplace), rather than as an educational aide, helpful framework, or accelerator? If so, can I live with the risks and consequences?

Summary

- AI systems are here to stay. Educators need to make peace with this, students already have.

- There are currently little-to-no legal frameworks in most countries relating to academia to effectively manage, limit, or guide the use of AI. The exception, which goes beyond AI, are those generally already in place to prevent outright cheating/plagiarism.

- Educators should recognize that many students will need AI skills in their careers, in the same way that basic literacy, numeracy, and tech-capability skills are already often required. Critical analysis skills relating to AI outputs are likely to be a core professional skill in the future.

- Students can fall foul of legal and ethical frameworks if AI is used in a way which is similar to existing methods of plagiarism, and sources are not verified.

- Students may be tempted to accelerate and facilitate their work using AI systems, but should be conscious of not shortchanging their own ability to solve problems and critically analyze AI outputs.

CHAPTER 3

AI for Marketing and Sales

The early functions of AI were limited in scope and rudimentary in capability. This rapidly changed. As computing power increased and data became more available, the potential for AI to accelerate and transform tangible aspects of real business functions, such as sales and marketing, became more practical. In the 21st century, the advent of big data and advancements in deep learning paved the way for sophisticated AI systems which are now at a level of accessibility that they are either cheap or free. By contrast, regulatory frameworks to mitigate or limit the practical applications of AI for business purposes are early-stage (at best).

Before the massive AI-wakening with the launch of ChatGPT in late 2021, tech giants had already been harvesting huge amounts of data gathered from daily personal and business usage of the Internet.[1] This data is often leveraged to generate product advancements and development with a customer-focus. Retention and acquisition of customers being the obvious goal. In addition, this data may be categorized and packaged as useful outputs for third parties to purchase so that they can either review it in isolation or combine it with additional data to understand their own existing and target customers.

---

[1] Vigderman, A. and Turner, G. (2024). *The Data Big Tech Companies Have On You*. [online] Available at: `www.security.org/resources/data-tech-companies-have/#:~:text=They%20also%20collect%20your%20name`.

© Harry Borovick 2024
H. Borovick, *AI and the Law*, https://doi.org/10.1007/979-8-8688-0400-7_3

Until the early 2000s, the cost of customer acquisition for many businesses included marketing costs as a whole, but would not have generally included tech usage data. Now, marketing budgets will often factor in data acquisition and targeting advertising using that data. In the past, this would have been likely considered a product-related cost, if considered at all. In part, this shift is because of legal frameworks.[2] The law in many jurisdictions, including across the UK, EU, and North America, has created many protections for personal data. Accordingly, it has become much more expensive for companies to legitimately acquire and utilize consumer data. These protections, hoops to jump through, also extend to the personal information of individuals within companies – to apply to B2B as well as B2C marketing and sales approaches.[3] However, with the rise of search engines in the late 1990s, and high-speed Internet in the early 2000s, the inclusion of data-related costs within marketing budgets became the norm. Now, it would be uncommon, likely an active choice, for even small businesses to not start their marketing approach with strategies such as search engine optimization (SEO) or targeted advertising based on cookies (or similar technologies based on tracking consumer habits). In theory, these marketing strategies should be conceived and carried out in compliance with an increasingly diverse and interwoven set of different data, privacy and marketing laws across jurisdictions. The outputs of SEO and targeted advertising both rely on complex algorithms. Early iterations of these algorithms are some of the first which were applied to a mass-market and directly impacted the daily habits of the general public. The opportunities to harness algorithms, now within the framework of AI systems, for both marketing and sales teams have only grown.

---

[2] Dunn, D. (2022). *How has GDPR affected Marketing & Data collection? - paperplanes.* [online] Available at: https://paperplanes.co.uk/how-has-gdpr-affected-marketing/ [Accessed 24 May 2024].

[3] Direct marketing detailed guidance Information Commissioner's Office. (2022). Available at: https://ico.org.uk/media/for-organisations/direct-marketing-guidance-and-resources/direct-marketing-guidance-1-0.pdf.

The Importance of Increasing Efficiencies

Marketers and sales professionals have been harnessing (to varying degrees of sophistication and success) AI's analytical prowess for customer insights, personalized targeting, and automated decision-making.

Mass adoption has already taken place – recommendation engines and predictive analytics have reshaped the marketing and sales strategies of businesses in almost every sector and challenged traditional approaches. Or, at least put a spotlight on more traditional marketing and sales methodologies to justify their return on investment. This is likely to continue to develop as SEO morphs into AIO, whereby companies optimize their online presence and commercial arrangements to be more inclusive within the answers given by AI systems. For example, when a consumer asks ChatGPT "Where can I buy a car near me?" – being a part of the natural-language response from the chatbot may become even more valuable to a dealership than simply appearing high up on a Google search listing. Even Google itself is moving toward an increasingly "question and answer" model, rather than simply producing a list of websites.[4]

However, as businesses attempt to navigate the available short-to-medium term opportunities from using AI (or even, much less sophisticated tools), it's important to remember that efficiencies and opportunities need to be balanced against potential pitfalls and avoid compromising long-term gains. The blending of artificial intelligence with the "art" of selling and promoting products and services is an art in its own right – there is no single right way, requiring a nuanced approach.

---

[1] Vallance, C. (2024) *Google using AI to come up with search answers in UK trial.* www.bbc.co.uk/news/technology-68730138.

The Risks of Automated Decisions

The automated decision-making or automated processes which arise through the use of AI systems (even with a human action at the end) come with inherent risks. These risks can be mitigated most effectively by a strong understanding of the fact that they actually exist. This is amplified when scaling a team, which is only as strong as its weakest (or least aware) link. So, this means that training, education, and transparency are inherently important to minimizing the potential pitfalls of using AI tools for sales and marketing more broadly.

Excessive Reliance and Loss of Skills

The risks to the users of AI systems for automated decisions and processes are most likely to be greatest when they don't understand the rationale for the automation or place excessive reliance on an AI-driven process without an ongoing critical assessment of outputs and outcomes. This is related but distinct from the issue of insufficient transparency, though it may arise in significant part because of opaqueness in the selection and application of AI systems.

If, for example, a user relies on AI tools to help them draft copy-text for adverts, they may see immediate and dramatic time savings. However, if they do not then gauge the quality of the results (e.g., the success of an advertising campaign, such as the sales of a product) against their prompts/inputs, it is unlikely that they can say with confidence the AI tools are actually improving their overall workflow. Speed and minimum effort are not always the same as overall satisfaction with outcomes. This is particularly important when we bring legal and ethical risk considerations into the equation.

In addition to taking the time to apply their own critical analysis, if a marketer is not actively seeking and assessing the feedback on their content (most likely to occur when they are producing content as a

54

subcontractor for a third party), they are less likely to accurately gauge the success and positive reactions to that output. More seriously, by not actively seeking that feedback or putting in place active checkpoints, they may be missing or may misunderstand the severity or nuances of potential negative feedback/displeasure by their own customer, internal stakeholders, or the intended audience.

Failure to understand, and if necessary address, negative feedback on marketing can materially increase the potential risk of being exposed to a legal action for negligence, misuse of IP, breach of contract or breach of laws – for example (as further discussed in the following) when an advert may be misleading, defamatory, or created on the basis of inappropriately acquired data.[5] The risks go even beyond the content itself, but how it is created and how it may be delivered (e.g., direct marketing) which may have their own set of legal restrictions and considerations.

The same applies to the use of AI tools in sales, where the short-term commercial risk is tangible, that is, the loss of a deal, and the serious longer-term legal risk and ethical concern of mis-selling or misrepresenting during a deal. Ethical considerations aside (and these can be significant), this could have severe legal and financial impacts on a business which gets its strategy and implementation wrong.

Lack of Agility to Navigate Reputational and Legal Risk

The behavior and demands of consumers and businesses can shift rapidly. The messaging content and approach which will resonate with consumers and businesses may also diverge over time. This is why regular iteration

---

[5] Farnsworth, N., Colt, A. and Hagedorn, K. (2023) *Managing legal risk in Marketing and Advertising: 5 things Companies should know.* www.orrick.com/en/Insights/2023/01/Managing-Legal-Risk-in-Marketing-and-Advertising-5-Things-Companies-Should-Know.

and pivoting is often required for brands to gain and retain relevance.[6] For example, consumer-facing marketing may trend so as to become increasingly eye-catching, while business-to-business marketing may remain more straightforward or based on the potential for a product or service to deliver some kind of tangible value.[7] Over time, what is legally acceptable (or ethically tolerable) may also shift. Laws and guidelines regarding marketing best practice within the bounds of local laws, for example, those issued by the ASA in the UK or the FTC in the United States, develop periodically. Laws and enforcement regarding the advertising of specific products can experience sudden shifts or restrictions. Alcohol and tobacco products being obvious examples of previously unrestricted advertising which have received either partial or total bans, or at least significant caveats. Marketing and sales strategies which are aimed at children, may impact children, or use the data of children[8] also attract specific restrictions of some kind in most countries.[9, 10]

---

[6] Pilcher, L. (2024). *The Rapidly Changing Digital Marketing Landscape: Why Marketers Remain Crucial.* [online] Pilcher Creative Media. Available at: www.pilchermedia.com/the-rapidly-changing-digital-marketing-landscape-why-marketers-remain-crucial/.

[7] Wenger, S. (2021). *B2B vs B2C Marketing – Differences and Similarities.* [online] B2B Marketing World. Available at: www.b2bmarketingworld.com/definition/b2b-vs-b2c/.

[8] ico.org.uk. (2023). *What if we want to target children with marketing?* [online] Available at: https://ico.org.uk/for-organisations/uk-gdpr-guidance-and-resources/childrens-information/children-and-the-uk-gdpr/what-if-we-want-to-target-children-with-marketing/.

[9] Federal Trade Commission. (2024). *Children.* [online] Available at: www.ftc.gov/business-guidance/advertising-marketing/children.

[10] Practice, A.S.A. | C. of A. (2024). *Children: Targeting.* [online] www.asa.org.uk. Available at: www.asa.org.uk/advice-online/children-targeting.html#:~:text=Rule%205.1%20states%20that%20whenever.

AI systems generally rely heavily on historical data. Dependency on historical data, without updated context or correction by people is far more likely to result in higher risk, and probably less effective, marketing or sales strategies. Picture a 1960s advert for cigarettes – marketed as being "cool." Through the lens of a modern eye, these adverts if released now would not only be considered ethically questionable and ineffective, but certainly illegal in many countries. By contrast, alcohol advertising remains legal in Europe, the UK, and the United States, but all within specific frameworks. Often, such adverts include disclaimers or are only advertised in specific locations or within restricted times. Without context, tailoring, and a localized approach, a "one size fits all" strategy suggested by AI systems would be almost certain to fail in a multitude of ways – in the same way that the strategy would fail if blindly proposed by an inexperienced or poorly researched/advised person.

So, any increased inflexibility to a business in creating, selecting, approving, or releasing marketing as a result of deeply integrating AI systems is a potential risk. The risk is exacerbated by where the business may have fallen into the trap of excess reliance. This can be mitigated. If a business (or even any individual trying to market/sell something) relies too heavily on AI systems in such a way that their process is entirely dependent on one or more AI systems, it needs to be recognized that the AI systems may not adapt quickly enough to evolving trends. This can make it hard to maintain a perception by the buyer of a human connection – something that many people actively want to feel when making a buying decision.

Although many business are dependent on specific technologies, the AI space benefits from rich competition of system providers, all pushing hard to have the latest and greatest innovations. So, it will often make sense for businesses to either diversify their suppliers or have a clear understanding of who their redundancy suppliers might be. For example, if a marketing designer uses an image generator such as MidJourney, it would be prudent to have a working understanding and a business continuity strategy which also incorporates another provider such as

DALL-E.[11] Often, using multiple systems in parallel will simply allow the human user to verify that they are not unduly limited in their outputs by how one AI system "thinks" vs. another. Changes to legal frameworks applicable in particular jurisdictions (e.g., around intellectual property law) may also impact particular suppliers, meaning they become more limited in their capability or quality. While the advice "invest more, use multiple suppliers" may sound like a very simplistic advice – it is hard to beat in practice.

It is also worth remembering that when utilizing AI systems to make decisions or create output for marketing and sales, many responsible AI providers allow their users to add significant conditions to the system's outputs. In addition, it is often possible to somewhat interrogate an AI system (particularly when text-based or multimedia such as ChatGPT-4o) and explain it to justify or explain outputs.

Excessive reliance on any technological aide, including AI systems, could tangibly reduce the effectiveness of marketing, reduce sales as a result, and further impact the overall sales strategy. It's arguable that the reliance on AI systems will become more entrenched than most technologies, as AI systems' "magic" is their ability to help in solving a problem without the user quite knowing what is going on inside the system.

This is often called the "black box" problem. It basically means that people don't know how an AI system has worked, or came to a conclusion/ given a particular output.[12] However, unlike with people, AI systems are likely to become more transparent over time. The "black box" is likely to be a symptom of the early stages of the AI boom, rather than a long-term

---

[11] Guinness, H. (2024) *Midjourney vs. DALL·E 3: Which image generator is better? [2024]*. https://zapier.com/blog/midjourney-vs-dalle/.

[12] Bagchi, S. (2023). *Why We Need to See Inside AI's Black Box.* [online] Scientific American. Available at: www.scientificamerican.com/article/why-we-need-to-see-inside-ais-black-box/.

problem for ordinary users.[13] The black box problem may also apply to a human being. We can never truly understand why a person has given us a particular answer, nor their mental process in creating something (e.g., their thought process while making an illustration for an advert). Nevertheless, people may often be more trusting of some information we are presented by technology compared to when the same information may be presented by people – increasing the chances of excessive reliance over time. This may be compounded when considering that, over a long enough period, for example, after the 50th advert in a long-running series of adverts, it may become increasingly practically difficult to understand and recall whether, and to what extent, the narrative of that series of adverts was impacted in particular ways by the outputs of AI systems – probably increasing the reliance and dependence on AI systems over time.

Practically, for sales and marketing, this means that users may gradually become uncertain about what decisions, suggestions, or approaches are their own and which have been influenced to some extent by AI. A greater concern still is that users may lose the skills, capability, or resources to generate non-AI-derived alternatives. This could be particularly catastrophic where a legal issue has arisen such as the AI system recommending something (probably innocuously, without negative intent) which could be simply illegal or give rise to other civil legal issues. While AI systems don't really (yet) have "intent," they can simply be wrong or mislead[14] or misuse something which belongs to someone else.

---

[13] Ropek, L. (2024). *New Anthropic Research Sheds Light on AI's "Black Box."* [online] Available at: https://gizmodo.com/new-anthropic-research-sheds-light-on-ais-black-box-1851491333?_gl=1 [Accessed 25 May 2024].

[14] Williams, R. (2024) "AI systems are getting better at tricking us," *MIT Technology Review,* 10 May. www.technologyreview.com/2024/05/10/1092293/ai-systems-are-getting-better-at-tricking-us/#:~:text=Ialk%20of%20deceiving%20humans%20might,users'%20expectations%20and%20feel%20deceitful.

The Consequences of Legal and Reputational Risk

It is worth considering, as it will arise throughout this book, that not every legal risk, nor everything which can give rise to a legal claim, is *illegal*. The point of this book is to have a legal eye on various complex issues, rather than explain the detail and nuances of the law. With that in mind, it is helpful to broadly understand

1. **Illegality (enforcement by authority)**: There are rules and laws which can be breached through actions, or even intentions behind some non-illegal actions. If the law is breached, there doesn't necessarily need to be a victim of any kind for a regulator or relevant body, perhaps even the police, to take action against the accused perpetrator. **Example**: An advertising regulator, such as the ASA in the UK, issues a direct ban on an advert by a company, which the regulator has determined is not compliant with the law.

2. **Civil infringement (enforcement by aggrieved party)**: Most countries have legal mechanisms and principles under which one party can bring legal claims or actions of some kind against another party. Often, this is called a civil "claim" or "suit." These actions will typically touch, or be based on, a field of law, but not be illegal *per se.* **Example**: One party owns some intellectual property (founded in relevant jurisdiction-specific laws which protect that IP) and sues another party for misuse of their protected IP to either stop the misuse or claim compensation.

The penalties (fines by regulators, or direct lawsuits from individuals and businesses who feel wronged) for misleading advertising can be severe.[15] Irrespective of truthfulness, penalties can often be even greater, in practice, where the advert is delivered or targeted or delivered in a manner which may not be in compliance with the law. The financial impact on a business of having to withdraw adverts which have already had significant effort and financial resources injected can be equally catastrophic.

Regulator-directed withdrawal of adverts can have direct costs, significant wasted costs, as well as unquantifiable reputational harm which can actually go directly against the originally intended messaging, potentially leaving companies worse off than if they had not advertised at all.[16] These potential risks apply to businesses which carry out marketing or direct sales strategies, irrespective of the involvement or application of AI. The overall risk is only materially increased when the involvement of AI is either poorly considered, or when it becomes entrenched, difficult to remove or bypass.

However, AI systems may touch or be used within marketing or sales strategies, the legal risks grow when AI (as will any technologies) are integrated into practical processes in such a way that either

1. The outputs make it to an audience while fully automated, without supervision/sign-off or

2. The intended outcome of using the AI system is potentially illegal by design e.g. intentional manipulation of people's decisions or exploitation of personal vulnerabilities.

---

[15] *Notices of penalty offenses* (2023). www.ftc.gov/enforcement/penalty-offenses.

[16] www.campaignlive.co.uk. (2002). *Libby's forced to withdraw natural ad claims by ASA*. [online] Available at: www.campaignlive.co.uk/article/libbys-forced-withdraw-natural-ad-claims-asa/80747 [Accessed 25 May 2024].

It is forbidden to use AI with sneaky or subconscious methods to manipulate customer behaviour. That includes for example posting fake reviews or testimonials from fake customers. Another thing that is prohibited is using AI to target specific demographic groups based on sensitive attributes like their race or religion.

—Tom Staelens, Upperscore[17]

The potential risks are not just legal. Misleading or defamatory marketing or sales strategies can also erode the trust by potential or current customers. Law and ethics don't always overlap, but it is more common than not for there to be a convergence. It's unlikely that if a company has potentially lied to, or somewhat misled, its customers (however intentional or not), that this won't have negative impacts on sales to customers and staff morale.

Even in circumstances where a regulator or authority decides there are insufficient grounds to bring a legal enforcement action, it may still be particularly challenging for a company or brand reputation if the matter becomes a publicly discussed matter with media coverage. Some media outlets even actively review notices of rulings by advertising regulators in order to find news content, which can often "create a second wave of adverse publicity."[18]

A key example is that of mass direct email tools, commonly used for marketing outreach and direct sales. This is one of the most common strategies across sales-marketing functions. It is a common complaint that

---

[17] Staelens, T. (2023). What marketers should know about the European AI Act. [online] Upperscore. Available at: www.upperscore.be/what-marketers-should-know-about-the-european-ai-act/ [Accessed 27 May 2024].

[18] Alderson, L. (2020) "The Advertising Standards Authority is investigating a complaint that our advertising is misleading. What should we do?," *Clarion*, 14 September. www.clarionsolicitors.com/articles/regulatory-asa-investigating-a-complaint-that-advertising-is-misleading.

such emails can feel very annoying to customers or prospects when poorly executed. Or, simply that many people receive these in such great volume that they can become a collective irritation. The downside of this is that they are often ignored even when they might actually be of interest to the recipient, were they inclined to engage. AI systems are already common which build upon existing systems for mass email outreach. Reliance on such AI tools to accelerate mass emails has the potential for great returns, in the right circumstances. However, if the mechanism is ineffectively applied, adding AI is unlikely to increase the changes of positive net outcomes.

Such tools are far more likely, when compared to organic person-to-person email or social media outreach, to fall on the wrong side of data privacy, marketing, and ecommerce regulations in many jurisdictions. Typically, this is because of where the contact details for the people receiving the outreach are collected. If appropriate consent is in place, as many providers of data claim, then risk is lower. Many countries have laws which, by design, have the effect of equating greater volume and lower personalization with less probability of legal compliance.[19] At the very least, mass direct marketing communications should be approached with greater care than either (1) direct but tailored, personal communications; or (2) marketing which is not directly sent or made out to individuals.[20] In any event, some feeling of personalization typically increases the percentage of success in this type of wide-reach marketing.[21]

---

[19] www.lawdonut.co.uk. (2024). Your email marketing and anti-spam law. [online] Available at: www.lawdonut.co.uk/business/marketing-and-selling/marketing-and-advertising/your-email-marketing-and-anti-spam-law.

[20] GOV.UK (2012). Marketing and advertising: The Law. [online] GOV.UK. Available at: www.gov.uk/marketing-advertising-law/direct-marketing.

[21] Marksons, S. (2023) "The role of personalization in effective B2B sales outreach," Medium, 4 October. https://medium.com/@sarahmarksons23/the-role-of-personalization-in-effective-b2b-sales-outreach-c1aea86d712a.

In short – it makes sense to start from a position of "should we?" and then "*how* should we?" rather than just "can we?" when deciding whether using AI tools for sales and marketing is the right approach. When AI systems are already in place, it is often just as important to consider whether existing tools are actually proving useful (and how), before deciding to just add more to your technology stack. This is likely to leave you or your organization sufficiently agile to avoid reputational risks in sales and marketing efforts, while also making the most of technologies which actually accelerate business without incurring significant potential commercial and legal downsides. If in doubt, it often also makes sense to engage a lawyer, if your business does not have its own, to review the terms and conditions of your preferred AI suppliers in order to understand how they collect the data that they may then provide to you. After all, if you use that data for your sales and marketing, you are likely to be the visible party to a marketing recipient and face legal, the immediate legal, and reputational consequences.

Defamation

Defamation is a type of legal action which may arise when one person says (typically called "slander") or writes (known as "libel" in many jurisdictions) something which is untrue or partially (but materially) untrue about another person. Crucially, that statement, whether written or spoken, needs to have actual potential to cause serious harm of some kind and/or financial loss. It's also possible for a company or group to defame or be defamed. The person who feels they have been defamed typically raises a lawsuit seeking some kind of (typically financial) compensation from the person who they believe caused their hurt. It may be that instead of money, the party which feels they have been hurt seeks some kind of retraction, apology, or other injunction[22] (a legal action which is either restrictive or forces an action[23]). There are many defenses to defamation, one of the most important being "truth,"[24] that is, that what has been said or published is substantively true, even if it is potentially harmful.

As a user of AI systems, it's important to consider whether there is excessive trust in outputs. When using AI systems to help create copy-text or images, users should consider verifying the truth of what is being generated. It is the visible creator and/or publisher (i.e., the company marketing using that copy) who is likely to be held primarily legally responsible for what they put out into the world. Complexities arise when secondary sources of information may also be potentially unreliable (e.g., using a manual Google search and the information on unverified websites to crosscheck AI outputs). So, practically, it will be sensible to carry out regularly evaluations of the potential for the outputs to harm others either financially or in some other material way (such as damage to reputation). A sensible approach to this type of evaluation is to start with looking at the outputs, such as an advertising slogan or messaging, on their own merits as a potential source of distress and harm – as well as within the context of its truth.

At the very least, it seems sensible to weigh up whether any third party might consider the text or image to be hurtful, stressful, distressing, or rude when read or seen in the coldest light of day and the harshest possible tone. Practically, it's very hard to avoid writing text or creating an image to please and avoid the displeasure of everyone and anyone. So, when putting together AI-generated or AI-assisted outputs, or verifying the outputs of AI – using the

---

[22] Foss, R. and Mowat, R. (no date) Injunctions FAQs. www.kingsleynapley.co.uk/services/department/dispute-resolution/injunctions/injunctions-faqs (Accessed: May 27, 2024).

[23] "injunction" (2024). https://dictionary.cambridge.org/dictionary/english/injunction.

[24] Legislation.gov.uk. (2013). Defamation Act 2013. [online] Available at: www.legislation.gov.uk/ukpga/2013/26/crossheading/defences/enacted.

truth as the guiding principle is likely to mean that marketers and salespeople are likely to minimize their legal risks – even if some risk cannot be entirely avoided. It should not be significant news or groundbreaking to anyone in marketing or sales that double-checking the truth or the potential to offend within something before it's published, or said, to potential or existing customers, is likely to be a sensible part of your processes.

Misrepresentation and Mis-selling

The words "misrepresentation" or "mis-selling" may be used casually as descriptive terms but have strict legal meanings depending on the country or region in which they are used. They may also have specific additional meanings or implications depending on the product or service to which they attach (e.g., financial services typically have specific and stricter rules).[25] Both apply to most types of marketing/advertising, as well as direct sales strategies.

Misrepresentation prior to, or during, a transactional negotiation, can vary in severity and potential liability. While there can be significant legal penalties, the risks are typically most severe where the recipient is an individual consumer or a group of consumers. Consequences are typically far greater, irrespective of whether the victim of the misrepresentation is a business or a consumer, when it is considered by a court to be fraudulent, rather than simply negligent.

Once a deal is actually done or money has changed hands, legal consequences for misrepresentation typically also increase.[26] This is because, in many cases, a practical or tangible loss has actually been suffered by the person who has been on the receiving end of

---

[25] Financial services mis-selling: regulation and redress Forty-first Report of Session 2015–16. (2016). Available at: https://publications.parliament.uk/pa/cm201516/cmselect/cmpubacc/847/847.pdf.

[26] Hall Ellis Solicitors (2019). Misrepresentation: Negligent & Innocent statements in contract law. [online] Hall Ellis Solicitors. Available at: https://hallellis.co.uk/misrepresentation-negligent-innocent-fraud/.

the misrepresentation, and that misrepresentation was some kind of inducement to make the deal happen. Even if no real loss has occurred, or no deal is done on the basis of the misrepresentation, it would still be potentially unethical and the basis of some types of legal claim. But, it's important to be clear that there is a formal meaning to misrepresentation which is distinct from mis-selling. The reason it matters is that in some specific circumstances, for example, in the case of an advert which doesn't even result in a sale but is false/inaccurate and aimed at ordinary consumers, there may still be potential civil or criminal consequences.

Whereas mis-selling in many jurisdictions typically requires an actual deal or exchange of money/goods/services. Often, it is falsely believed that someone actually has to suffer a real loss (i.e., mis=selling rather than misrepresenting) for there to be significant risks to the advertiser or the seller. In some countries, often depending on who the potential audience/buyer might be – there is little to no distinction between misrepresentation and mis-selling.

This means that, advertisers and sellers need to remember that they may be legally responsible for the financial, emotional, reputation, or other harm felt and experienced by a recipient of their marketing or the buyer of their goods, if that recipient/buyer feels misled. The fact that the misleading act, statement, or idea was generated, inspired, supported, published, or targeted by AI systems the advertiser or seller has employed is unlikely to be any kind of reasonable defense.

The lack of a defense to legal claims, on the basis of innocent reliance on AI systems, may change as the law evolves in future, but for now – there needs to be a person or a company run by people to take responsibility and feel potential consequences. This is societally important as it means that marketing strategies and sales tactics, to stay on the right side of the law, need to stay relatively fair and truthful and therefore some kind of consumer and business protection is baked in. The importance of this is multiplied when applied to high-risk or more regulated products such as financial services. In early 2024, the US Federal Trade Commission (FTC) explicitly stated that

... there is no AI exemption from the laws on the books[27]

Importantly, in many countries, even if you have a contract with the person who bought your product or service, you might not be able to exclude your liability to them irrespective of what the terms of a contract might say.[28] The result could be that financial compensation or other legal remedies are available to them, or they can report you to a regulator who could issue fines or other penalties, particularly if there is the potential that you have acted in a manner which could be considered criminal by nature or severity.

"Buyer beware" or "buy at your own risk" often only goes so far, particularly when selling to ordinary consumers who benefit from additional protections compared to businesses in many countries. There are legislative protections (laws) targeted at providing protection specifically to consumers, where marketing is aimed at members of the public rather than businesses. In the UK, the most prominent example is the Consumer Rights Act (2015).[29] In addition, regulators will be conscious and encouraged by government policy to treat business-to-consumer marketing under more onerous expectations, presuming that consumers are more vulnerable than businesses.

For example, in the UK, the Advertising Standards Agency (ASA) has specific codes and guidance for marketers and salespeople when targeting direct-to-consumer advertising or sales-pitches. As does the UK Competition Markets Authority (CMA) which aims to prevent anti-competitive (known as "anti-trust" in the United States) practices by

---

[27] Consumer Facing Applications: A Quote Book from the Tech Summit on AI (2024). www.ftc.gov/policy/advocacy-research/tech-at-ftc/2024/04/consumer-facing-applications-quote-book-tech-summit-ai.

[28] Pease, C. and Parkinson-Cole, K. (2013) Illegality in contracts. www.inhouselawyer.co.uk/legal-briefing/illegality-in-contracts/. [Accessed on 30 May 2024)

[29] GOV.UK (2015). Consumer Rights Act 2015. [online] Legislation.gov.uk. Available at: www.legislation.gov.uk/ukpga/2015/15/contents.

companies.[30] The UK's Information Commissioner Office (ICO), which aims to safeguard the reasonable use of individuals' personal data,[31] also has similar and arguably stronger guidance (due to its basis in strict regulation – UK and EU GDPR) to protect the data of ordinary consumers and how that data can be used to reach out, target, or sell to those consumers. With this in mind, and calling back to the messaging put out by the US FTC, when using AI for sales and marketing "I trusted the AI" is unlikely to be considered a sufficient defense or excuse by any competent regulator (not just those discussed in this chapter) or court.

Why Misrepresenting, Mis-selling, or Misusing Consumers' Data Matters

Regulators and courts are typically much more harsh and less forgiving when the actual or potential victim of any civil or criminal act is an ordinary member of the public rather than a business. The law expects businesses to seek appropriate and proportionate legal advice, take a critical approach to what they buy and ask questions to gain comfort or assurances before making a commercial deal.

Consumers are not absolved of responsibility. They are expected within the regimes of most modern legal systems to still make somewhat rational decisions and to try to educate themselves regarding what it is they are buying. However, it is quite easy for even a layperson to understand and expect that another layperson, or even a very educated and shrewd person who is a member of the public, to have less resources

---

[30] GOV.UK (2019) Consumer vulnerability: challenges and potential solutions. www.gov.uk/government/publications/consumer-vulnerability-challenges-and-potential-solutions/consumer-vulnerability-challenges-and-potential-solutions.

[31] ico.org.uk. (2022). Our purpose. [online] Available at: https://ico.org.uk/about-the-ico/our-information/our-strategies-and-plans/ico25-strategic-plan/our-purpose/.

and negotiating power than a business or company of most sorts. The laws of most countries not only recognize this but, in many cases, impose significant additional rules and punishments onto businesses selling to consumers ("B2C") compared to when selling to other businesses ("B2B").

In 2019, the UK's Financial Conduct Authority (FCA) fined Carphone Warehouse, a major retailer of mobile phones, around £29 million for mis-selling one of its aftersales care products called "Geek Squad."[32] This example helps highlight how nuanced mis-selling can be. Whereas misrepresentation is often mostly easily understood as outright lies or half-truths in order to make a sale, mis-selling in this case was to sell a product which, as the FCA described "...*in some cases had little to no value because the customer already had insurance cover.*" [33] This is a prime example of a common situation where a product or service has been sold, is not an outright con, falsehood, or sham but is being sold as *knowingly unnecessary*. Accordingly, being simply unnecessary and yet strongly encouraged by sales teams to consumers, who are not expected by law to know better, is sufficient to be considered as mis-selling[34] and attract significant repercussions.

---

[32] FCA fines The Carphone Warehouse over £29m for insurance mis-selling (2019). www.fca.org.uk/fca-fines-the-carphone-warehouse-over-29-million-for-insurance-misselling#:~:text=The%20Financial%20Conduct%20Authority%20(FCA,which%20stemmed%20from%20whistleblowing%20reports.

[33] FCA fines The Carphone Warehouse over £29m for insurance mis-selling (2019). www.fca.org.uk/fca-fines-the-carphone-warehouse-over-29-million-for-insurance-misselling#:~:text=The%20Financial%20Conduct%20Authority%20(FCA,which%20stemmed%20from%20whistleblowing%20reports.

[34] Financial Ombudsman. (n.d.). Insight in depth: underinsurance, misrepresentation and non-disclosure. [online] Available at: www.financial-ombudsman.org.uk/data-insight/insight/insight-in-depth-underinsurance-misrepresentation-non-disclosure.

In this particular case, it didn't help the position of Carphone Warehouse that this type of aftersales care product/service is often regulated in the UK as a type of insurance product and therefore attracts extra scrutiny.[35] Although this situation happened without the involvement or reliance on AI systems, the lessons from this example illustrate a potential clear legal and reputational risk when relying on AI systems. Particularly, when those systems or tools lack greater context or legal guardrails. A prompt to an AI system asking it to help a company design or market an aftersales product is likely to yield exactly that – a helpful sales strategy or marketing materials, without the nuanced understanding of how carefully this output needs to be considered within the regulatory framework of the specific product, nor how different regions may approach the requirements of consumer protection. If anything, greater reliance on AI systems, which generate increasingly compelling and (in many regards) high quality output, is likely to lead to greater trust and less verification.[36]

So, much of the governance and raison d'être of regional and international regulators is specifically to protect consumers, or members of the public even when they are not acting as consumers/buyers, from the messages and actions of businesses.[37] We also know that businesses also sue or make a variety of legal claims against each other, and these can be just as financially impactful. In some cases, where a company may be

---

[35] Kollewe, J. (2019) "Carphone Warehouse fined £29m for insurance mis-selling," The Guardian, 13 March. www.theguardian.com/business/2019/mar/13/carphone-warehouse-fined-29m-for-insurance-mis-selling.

[36] Lukowski, J. and expert panel (2024) "11 Risks to using AI in Marketing (And How to Mitigate Them)," Forbes, 2 May. www.forbes.com/sites/forbescommunicationscouncil/2024/04/25/11-risks-to-using-ai-in-marketing-and-how-to-mitigate-them/#:~:text=Using%20generative%20AI%20in%20marketing%20poses%20the%20risk%20of%20creating,values%2C%20potentially%20damaging%20its%20reputation.

[37] Stanley, M. (no date) Understanding regulation. www.regulation.org.uk/specifics-consumer_protection.html. (Accessed: June 24, 1AD)

openly mis-selling or misrepresenting in its marketing or sales strategy it may bear an even greater risk where the approach is agnostic, with exposure to both consumers and businesses.

It's also worth remembering that the personal data of individuals (and rules around the misuse of that personal data to contact and sell to individuals) is likely to include their contact details, email address, and phone numbers – both personal and professional. So, even if cold-calling or cold-emailing someone in their business capacity can bear some risks. These risks, practically, are likely to be greater and compounded when the recipient has asked explicitly to not be contacted and then they receive further communications. This is often a delicate balance because it would be easily arguable that direct marketing or outreach is more effective in creating sales, whereas it is far more complex and challenging from a legal and risk perspective compared to indirect sales and marketing, as the recipient isn't simply viewing an advert which is open to any viewer. The very fact they are being *targeted* spikes potential risk significantly.

The exposure to potential risks may improve in future when more advanced prompts and AI system models are able to consider such subtleties. However, this may also require the providers of AI systems to essentially verge on giving regulatory or legal advice (i.e., integrating legal perspectives and considerations by design) – a big potential cost and risk, rather than just giving more generic and less tailored outputs. For now, it is more important to realize as a user of AI systems that many are designed to give an answer or output to (nearly) any request, even if those answers or outputs are false, inaccurate, or misleading. This is often called "hallucination," a risk that has become so common and far-reaching in the risk it creates that, in its application to AI, was Cambridge University Dictionary's "word of the year."[38]

---

[38] University of Cambridge. (2023). Cambridge Dictionary names "Hallucinate" Word of the Year 2023. [online] Available at: `www.cam.ac.uk/research/news/cambridge-dictionary-names-hallucinate-word-of-the-year-2023`.

Potential Consequences of Using AI Systems Sales and Marketing Without Appropriate Verification and Safeguards

1. **B2B misrepresentation and mis-selling:**

    a. **Reputational damage**

    b. **Lawsuits from businesses**

    c. **Police or regulator investigations and penalties for fraud****

2. **B2C misrepresentation and mis-selling:**

    a. **Reputational damage**

    b. **Fines by regulators**

    c. **Compensation payments ordered by regulators to consumers**

    d. **Lawsuits from consumers**

    e. **Police or regulator investigations and penalties for fraud***

3. **Misuse of personal data for targeted selling:**

    a. **Reputational damage**

    b. **Fines by regulators (these can be particularly huge)**

    c. **Compensation payments to consumers (ordered by regulators)**

Consequences can be far-reaching, variable, and complex – these are just some, at a high level.

**Typically, only seen in the most serious and intentional cases*

Bias in AI for Marketing and Sales

Related to the risks of automated decision-making without checks and accountability, there is also a significant possibility within sales and marketing teams to miss the impact that bias may have on outcomes. This can bring positives and serious negatives, but in any event – is a real risk.

The most obvious positive outcome is that a user of AI tools for sales and marketing is likely to simply see a percentage increase in success. However, the negatives may well have serious legal impacts, irrespective of whether commercial success metrics (e.g., marketing reach or sales figures) go up or down.

AI tools may help a sales and marketing team select targets for direct sales outreach, arguably the most direct form of marketing. The same or similar AI tools may help with the drafting of communications. Automated selection of marketing viewership is often, and increasingly over the past decade, left to automation or algorithmic selection (just think about all those targeted adverts which feel *a bit too* accurate and targeted). Automated decision making which makes use of personal data, that is, anything identifiable about who they are, often bears inherent legal risks – particularly where this results in profiling of individuals.[39]

Many legal regimes, such as the GDPR in the UK and EU, legally require a human to be in the loop of decision-making – technically called "human intervention."[40] Basically, the law in the EU and UK generally

---

[39] ico.org.uk. (2023). What does the UK GDPR say about automated decision-making and profiling? [online] Available at: https://ico.org.uk/for-organisations/uk-gdpr-guidance-and-resources/individual-rights/automated-decision-making-and-profiling/what-does-the-uk-gdpr-say-about-automated-decision-making-and-profiling/.

[40] Intersoft Consulting (2013). Art. 22 GDPR – Automated individual decision-making, including profiling | General Data Protection Regulation (GDPR). [online] General Data Protection Regulation (GDPR). Available at: https://gdpr-info.eu/art-22-gdpr/.

requires a human to be involved in the process to oversee and make final or overarching decisions where an automated tool such as an AI system is automatically making decisions about who receives or benefits, and who does not, from any commercial decision or action. This applies to sales and marketing, and businesses should be careful with expecting that they can entirely delegate these functions to AI systems.

For several years, I was a lawyer focusing on the field of segmented, automated, and intelligently targeted digital marketing. I know that this type of targeting has reached great levels of sophistication, with AI being central to many such systems. Segmentation of the audience is at the core of modern digital marketing.[41] So much so that often the recipients of marketing may not even realize that they are seeing a targeted advert. Most people would not assume that, unlike an advert on social media, they may well be seeing a targeted advert or different advert to someone else when viewing streamed television content or watching *ordinary* television (most of which is now digital with easily adjusted advertising breaks). This can have potential to cause harm, even if not inherently illegal by nature, for example, where there are risks of potential social harms such as impacts on mental health.[42]

---

[41] Deloitte (2022). How to leverage AI in marketing: three ways to improve consumer experience. [online] Deloitte Slovenia. Available at: www2.deloitte.com/si/en/pages/strategy-operations/articles/AI-in-marketing.html.

[42] Wachter, S., Milano, S. and Mittelstadt, B. (2021) Targeted ads isolate and divide us even when they're not political. www.oii.ox.ac.uk/news-events/targeted-ads-isolate-and-divide-us-even-when-theyre-not-political/. (Accessed: June 1, 2024)

This comes with three clear practical and legal risks:

1. **Practical:**

   a. **Shutting down success:** The recipient of an advert may be less receptive or even actively upset if they find out they are being discreetly advertised to in a targeted manner. People tend to be less upset they are being sold to, when they know they are being sold to.

2. **Practical:**

   a. **Shortsighted selling:** The targeting system is focusing excessively on the information it may have or understand about the viewer, oversimplifying their desires. This could mean that aspirational selling is a missed opportunity (e.g., selling to people for their current wants/needs rather than the more strategic selling).

3. **Legal:**

   a. **Targeting, on the wrong side of the law:**

      i. Bias may result in marketing or sales strategies aimed narrowly at people with specific characteristics, profiles, or traits. Depending on which country or region in which the recipient is based, this might have civil or criminal legal consequences. For example, UK and EU regulations explicitly make it illegal in most commercial circumstances to use an individual's racial, ethnic, or health information to target them. In many cases, it may even be a breach of law to simply hold that information.

ii. At the most extreme end, it is also clearly an ethical and legal wrong (in just about most countries with democratic legal systems) to target individuals based on their race, gender, or characteristics for just about any practical commercial reason, though this still happens frequently – for example, when the UK government was specifically targeting minority groups with adverts and propaganda.[43]

A fundamental issue with AI systems in sales and marketing, or most tools which *generally* rely on applied algorithms with historical data, is that they focus on the past and the present, and on specific sets of data which are by their nature historic. Marketing and effective sales strategies are broadly about future value. The marketed image of how much a person might want to feel good in a high-quality pair of boots, when effectively conveyed, may be the factor that convinces them to buy instead of relying on their experience of an existing worn-out pair or purchasing a lower-quality item.

This isn't to bash AI systems' abilities to extrapolate from current wants and haves, into some understanding of future wants. However, it is common for bias of many kinds to leak into the outputs of AI systems due to their *understanding* being based on existing data rather than inventive thought.

---

[43] Shanti Das – www.theguardian.com/media/2023/aug/13/government-targeting-uk-minorities-with-social-media-ads-despite-facebook-ban#:~:text=In%202021%2C%20Facebook%20announced%20a,reach%20and%20exclude%20minority%20groups.

Key Considerations

- If you use AI systems to help you devise or produce an advert, or send out marketing of any kind, you are still responsible for the content of that marketing, the truth of it, and how fair and legal the content may be.

- When undertaking marketing or sales strategies on the basis of data, particularly if personal data is used, it is important to consider how the data was gathered and its accuracy.

- The data used should be legally gathered, processed, and stored, and the provider of the data and the AI system should make appropriate assurances (ideally, contractually).

- The law, and the AI systems' T&Cs, make it unlikely for the AI provider to have any liability – so the risk may be all on you or your company/employer.

- If you are checking your output before it is published/ marketed/used as a platform or strategy to sell, double-check it with a colleague or someone you trust where possible. Diversity of perspective and experience can be helpful in identifying risks. If in doubt, speak to a lawyer.

- If you do misrepresent a product or service to someone ("this product does X," when it actually does "Y"), this may also be considered mis-selling. In either case, you may have committed a somewhat fraudulent, or potentially fraudulent sale. This is likely to have some potential consequences, these might be severe and can be a mix of financial and reputational. So, bear in mind the consequences before you make a risky decision or take a risky approach.

- If you're wondering what to check – put yourself in the shoes of a potential viewer/reader and think:

  - Am I likely to be offended or feel discriminated against by seeing, receiving, or being excluded from this?

  - Are we being truthful about what the product or service does, and why someone might want or need it?

  - Are any implications, even if they are not hard promises, true?

  - If the statements or implications are not intended to be promises, would I be likely to easily understand that?

  - Is the marketing sales pitch I am viewing or reading related to a product or service which may have specific legal restrictions and risks, for example, product – alcohol/tobacco, or service – accounting/legal?

CHAPTER 4

AI and Your Money

The interaction between automation and personal finances has been increasing in pervasiveness and complexity since the 1980s when digitized financial trading really started to become the norm.[1] However, automation has, with an explosive pace in the past 15 years, also become the norm for personal finances. It is no longer just the banks, insurers, and large financial institutions who benefit from, are at risk from, and rely on automated digitized processes. Very little of the modern financial system exists outside of a fully digital ecosystem.[2] Increasingly, AI systems, or at least algorithmic decision-making systems (which may be based on complex rules, rather than machine learning), have become more powerful and increasingly accessible.

---

[1] Windmill Smart Solutions AG (2023) All about fintech: history, development, and future. www.windmill.digital/all-about-fintech-history-development-and-future/.

[2] UBS Editorial team (2023) history-of-digital-banking. www.ubs.com/ch/en/wealth-management/womens-wealth/magazine/articles/history-of-digital-banking.html.

© Harry Borovick 2024
H. Borovick, *AI and the Law*, https://doi.org/10.1007/979-8-8688-0400-7_4

The Core Issues

Assuming that the vast majority of us hold some form of bank account,[3] it is almost certain that some form of automated systems already influence or determine the following issues:

1. How much money you can withdraw or deposit

2. Where your money can be spent or transferred

3. The hoops you may need to jump through to access your money securely

4. Your line of credit and the terms of any loan

5. Whether you can buy a home or a car, or even the terms of a rental/lease

6. How you are taxed

7. Where and how your money is invested

8. To whom your financial situation is disclosed, that is, which third parties know all about your finances

What the increasing availability of AI systems means for each of these situations is not universal. There is significant global variation, and even great differences between financial matters. For example, "how you are taxed" will vary significantly even within the EU, or even based on what type of financial asset or instrument is being considered. At the same time, access to basic AI systems is becoming global and, in many cases, free to use. There are great variables on the outcomes of the eight issues identified

---

[3] World Bank Group (2018) "Financial inclusion on the rise, but gaps remain, global Findex database shows," World Bank, 23 April. www.worldbank.org/en/news/press-release/2018/04/19/financial-inclusion-on-the-rise-but-gaps-remain-global-findex-database-shows#:~:text=Globally%2C%2069%20percent%20of%20adults.%20%20%E2%80%8C.

(and a complex net of other inter-related issues) based on where you live, your personal financial situation, where the third party (e.g., the bank or the relevant asset) is based, and the legal system that may govern the situation.[4]

Necessity

Broadly, in any country where a bank account is a necessity for a reasonable quality of life or access to many of the requirements of day-to-day life, individuals have never had less financial privacy.[5] Ordinary living is entangled with financial transactions and disclosures to third parties, often far more than we are aware. A simple credit card transaction to buy groceries often involves multiple parties deciding whether that transaction can occur, for example, the bank, the credit provider, the credit rating agency, etc. – even whether the seller accepts that type of payment method.

This picture becomes even more complex when considering that each of these third parties has their own stakeholders/third parties who impact on them in their own supply chain. For example, in much of Africa, digital banking transactions primarily occur via mobile phone applications, mobile credit accounts, and direct financial remittance systems.[6] Therefore, individuals in these regions are far more reliant on slower, though effective, mobile network infrastructure compared to equivalent individuals

---

[4] World Bank Group (2016) Digital Financial Inclusion, World Bank. www.worldbank.org/en/topic/financialinclusion/publication/digital-financial-inclusion#:~:text=Digital%20financial%20services%20%E2%80%94%20including%20those,with%20some%20reaching%20significant%20scale.

[5] Anthony, N. (2023) Cato.Org/The Right to Financial Privacy. https://cato.org/policy-analysis/right-financial-privacy. (Accessed: June 6, 2024)

[6] Statista and Taylor, P. (2023) Mobile money transactions in Africa 2020–2022. www.statista.com/statistics/1139403/mobile-money-transactions-africa/#:~:text=In%20total%2C%20mobile%20money%20users,transactions%20from%20the%20previous%20year.

in Europe or the United States where there are numerous alternative banking and transfer processes (e.g., physical bank branches), as well as technological systems for accessing those processes (online banking via fiber-based or short-wave high-speed wireless signals like 5G).[7]

Privacy (or Lack Thereof)

Therefore, with a greater reliance on third parties to be able to live an ordinary life, there appears to be an inherent and growing reduction in personal privacy. This is not new and has received occasional bursts of concern in the press when aspects of digital banking and finance experience some more visible changes (e.g., when "Chip and Pin" cards became the norm in much of the credit and debit card-using world).[8] To some extent, this leads to a reduction in freedom – typically still within the bounds of the law. The inability to secure credit, for example, could severely limit the reasonable freedom of an individual to access something that could improve their quality of life. The law has afforded individuals protections from these risks even prior to current concerns around the impact of AI.

The introduction of the General Data Protection Regulations in 2017 (commonly known as "GDPR") placed significant limits on the ways in which companies could use the personal information of individuals.[9]

---

[7] Harrison, P.J. and Harrison, P.J. (2021) Top African challenger banks helping the unbanked through mobile services. https://thefintechtimes.com/top-african-challenger-banks-helping-the-unbanked-through-mobile-services/#:~:text=Around%2057%25%20of%20the%20population,the%20answer%20comes%20from%20smartphones.

[8] Burton, J. (2016) "Banks that spy on your private life to flog their deals," This Is Money, 24 February. www.thisismoney.co.uk/money/saving/article-3460933/Banks-spy-private-life-flog-deals-monitor-shopping-habits-eating-holdiays-know-planning-baby.html.

[9] Dolea, O. (2018) Open Banking and data protection: Friends or foes? - The Global Treasurer. www.theglobaltreasurer.com/2018/03/19/open-banking-and-data-protection-friends-or-foes/.

These limits serve to protect the rights of all EU (and UK) citizens. These rights have, in theory, carried into the more recent AI-driven banking era – meaning that the way in which companies, banks, and other financial institutions can use the data of an ordinary person is as protected as before.

Whether these rights and protections are actually being respected in practice will become increasingly apparent over time – though there have already been numerous cases indicating that the financial sector has some way to go before there is universal compliance with the "in principle" rights of individuals. Even when there has been enforcement for misuse of personal data, punishments have been small or generally ineffective.[10] Quite simply, the more we have all grown dependent on highly digital bank systems who demand and hold ever greater amounts of our personal data, the more significant the risk to each of us is that we lose our privacy or that our data can be misused – with little or no consequences for the financial sector when it goes wrong.

The AI-Information Age

In many ways, personal choice and access to quality advice has also continued to grow. Even now, many people (in lots of cases, quite rightly) depend on direct financial advisory services. Some financial advisors may get a commission, or be fully independent – but the diversity of financial advice and availability of financial advice at different price points has never been greater.

In March 2024, I spoke with David Hancock. He was formerly the Chief Operating Officer at Brevan Howard, one of the largest hedge funds in the world and a company which was able to effectively use technology and data for financial gains. He concisely summarized much of the holistic content of this chapter. Most notably, speaking regarding knowledge and information as a financial tool:

---

[10] Pegg, D. (2020) "Arron Banks's firm and Leave. EU face £135k fines over data misuse," The Guardian, 3 February. www.theguardian.com/uk-news/2018/nov/06/arron-banks-firm-and-leave-eu-face-135k-fine-over-data-misuse.

The use of AI in finance has developed rapidly over the last few years. It is mainly used to develop "Black Box" or algorithmic trading models. Many large investment managers use these sorts of models to allocate capital and resources. For the individual unregulated person however, the picture is somewhat different. Depending on their knowledge of finance they could use AI in several ways. The main use would be as a learning tool in order to understand some of the data and terminology of the financial markets and how assets move given certain conditions. It may also help in decision making by answering specific questions i.e "if the non farm payroll number is below 200,000 what will the yield of the 10 year treasury bond be in 3 months time ?". So it would provide an unregulated person with a useful tool to help them understand and navigate financial markets and products, but like any AI enquiry it should be used with care particularly if you are using it to make decisions regarding your finances.

In most economically developed nations, and even in many rapidly developing nations, finance professionals of most types are regulated. Financial advisors are often responsible for the quality of their advice to an individual. So, for example, if a mortgage broker were to knowingly give very poor financial advice to drive an individual to take a significantly worse mortgage than is otherwise available to them, the broker may face civil or even criminal legal penalties had they gained a personal benefit such as a commission. There may even be penalties where the broker is not knowingly giving bad advice but is simply negligent. This is long-established to *try* and protect the more vulnerable person – the ordinary framework investor. The closer you are to being a "professional" the less protection the law affords you. While an individual's knowledge can be taken into consideration, if a financial professional is engaged to provide financial advice and does so in a circumstance as in the preceding, it does not matter the level of knowledge the recipient possesses on the protection they face.

As AI systems increase in complexity, they are increasingly being used by financial professionals as well as ordinary individuals in a variety of ways to improve personal financial outcomes. At the most basic level, this could be asking an AI system such as ChatGPT or Bard to explain a financial product or service, but could go as far as asking such systems for actual financial advice. More complex systems are capable of actually placing/buying/selling financial investments or commitments in an automated way on the basis of the AI's suggestions. This comes with great opportunities and significant potential risks.

Leveraging AI for Greater Financial Opportunities

Disclaimer: It's important to say that nothing in this book is financial advice or a suggestion as to how you should save, invest, or spend your money. You should always carefully consider your financial position before making any significant action/change and seek financial advice which is (if at all possible) independent.

This disclaimer should not be read as a simple risk warning to glaze over – but is fundamental to understanding how you should think about the use of AI systems for managing your finances, as well as the views and advice of anyone who may try to steer you in a particular direction or down a particular course of action. Remember, AI systems are made by people, and people typically have some form of self-interest. This is not to say that AI systems cannot or should not be used by ordinary people to better understand the financial system or how money should be allocated, but that – just like when leaning on other people for the same things – it's generally advisable to interrogate what you are told and to sense-check with multiple sources. An important nuance is that AI systems may be carrying the bias of their creators or their source data, even if the primary purpose of the system is not to actively drive the user one way or another.

87

Broadly, the law is clear – when a person or a technological system provides financial advice or financial services to an individual (particularly where the individual is an ordinary person, that is, "non-sophisticated investor"), this should be done with due care, attention, and independence. Where there is reduced or absent independence, the motivations of the provider of the advice should be declared or understood by the end user. At the very least, in many jurisdictions, it is expected that some bias is declared even if the specifics are not entirely clear. For example, when taking advice from a mortgage broker, it should be clear that the mortgage broker may receive a commission from the bank which provides the loan, should you follow their advice. Or, if the person giving you the advice is also directly selling or profiting from your investment decision.

The Framework of Risk

It is important to understand the core of the concerns that providers of AI systems may have before looking at the risks and potential mitigations for ordinary users. Most modern legal systems have some legal frameworks and restrictions in financial services against

1. Negligence: You are typically owed some form of duty of care, or at least a clear and unambiguous disclaimer against any potential negligence, that is, "Use this at your own risk, the value of your investment may go down as well as up."

2. Mis-selling: Already discussed in Chapter 4, but often with greater risks and consequences when applied to financial advice, products, and services.

3. Unqualified advice: Someone or a system giving you financial advice when they are not reasonably qualified to do so.

Points 1 and 3 are more complex when considering AI systems that ordinary individuals may interact with, compared to the human alternative (e.g., a typical financial advisor). It is very hard for an AI system to appropriately determine when it is simply providing an educational service, or when it is providing the type of financial advice that it should be disclaiming against or be regulated and monitored when providing. It's far more reasonable to allow, expect, and support ordinary non-sophisticated investors in self-education.

Accordingly, providers of AI systems are at far more risk of legal challenges by users, and investigations by regulators, when attempting to answer a question such as

   a.   *"Should I invest in a mutual fund?"*

rather than

   b.   *"What is a mutual fund and how do people make*
         money investing in them?"

While the latter half of question b is arguably seeking investment advice, it is clearly more ambiguous and a less direct request for advice than question a. At the very least, it is clear that question b's primary objective is educational rather than to seek an explicit steer.

What is a notable and increasing trend, however, is the growth of specialist systems that are built and designed to actively bear greater risk with the specific goal of being market disruptors. These providers of more narrowly defined specialist AI systems are increasing in number.[11] They typically vary between those who pitch as providing higher-grade but non-regulated advice compared to those who are actively trying to demonstrate they are capable of providing high-quality financial advice. Numerous

---

[11] How Artificial Intelligence is Transforming the Financial Services Industry (2023). www.deloitte.com/ng/en/services/risk-advisory/services/how-artificial-intelligence-is-transforming-the-financial-services-industry.html.

banks have actively marketed that such systems are in the works to benefit their customers.[12] Whether such systems ever become particularly usable/useful or available to non-sophisticated investors remains to be seen.[13]

The Perspective of the User

It will likely remain the case for the foreseeable future in most jurisdictions that, a regulated financial institution providing financial advice or financial management via an AI system to another person with financial qualifications or experience, is a lower risk of catastrophic long-term loss to either party compared to a self-service investment tool that anyone can access and on which significant reliance may be placed.

Where there are few/no protections or limitations on who can access financial systems which are typically utilized by regulated professionals, anyone can use or directly lose money when using. This is a clear legal and regulatory risk that providers of AI tools and financial advice are typically keen to minimize. However, many jurisdictions have a more relaxed or legal approach to financial regulation, AI regulation, and investor controls.

So, in some countries which have historically been less concerned with digital consumer protections and have fewer banking restrictions (such as many parts of the Africa or South Asia), it is quite possible that we may see greater use of self-service AI-powered financial investment systems. This could be argued to mean that these jurisdictions will have more democratized wealth creation opportunities and risks compared to more highly regulated European or North American jurisdictions, at the cost of greater risk of personal financial loss without human regulated advisory

---

[12] Son, H. (2023) JPMorgan is developing a ChatGPT-like A.I. service that gives investment advice. www.cnbc.com/2023/05/25/jpmorgan-develops-ai-investment-advisor.html.

[13] Artificial Intelligence research (no date). https://jpmorgan.com/technology/artificial-intelligence.

and investment support. Crucially, for any user of AI-powered financial advisory, financial management, or investment tools – irrespective of the country in which the user resides – many of the key questions and risks remain the same:

1. What is the source data which sits within the engine of the AI and how reliable/current is it?

2. Is there a person or company with vested or conflicted interest running the tool?

3. How much of a cut, if anything, does the tool take if the user makes a gain? If nothing – how does the tool make money and does that affect its independence?

4. For advisory services, are taxes applicable in the jurisdiction of the user being factored in?

5. Is there a taxable gain to consider or pay as a result of a benefit out of using the tool (and do non-AI powered tools offer potentially more tax-advantageous options)?

Questions such as how to save, invest, retire, or borrow money or buy, leverage, and sell assets have always been complex. As complexity has increased, so has the varied availability and quality of professional advice. Financial advice is expensive. Some of the reasons for this expense are legal, such as the need for regulated professionals to have professional indemnity insurance. However, a combination of marketing and the reality that there is a limited supply of high-quality financial advisor both contribute significantly to an often prohibitive cost.[14]

---

[14] Neugarten, J. (2023) What fees do financial advisors charge? www.investopedia.com/ask/answers/091815/what-fees-do-financial-advisors-charge.asp.

Bypassing or minimizing this cost is very attractive for a potential user, particularly for individuals who may find *any* kind of professional financial advice too expensive. As AI tools can potentially negate the issue of supply, it is worth noting that quality and trustworthiness can vary wildly. While a progressive increase in quality of AI-based financial advice is highly likely, it will never be perfect. Just like the best human financial advisors are imperfect, there are simply too many variables for any system to be right every time for every person – particularly when accounting for personality, risk appetite, and personal circumstances. As AI systems improve, it will increasingly make sense for people to choose the cheaper of the imperfect systems, that is, specialist AI over the *average* financial advisor, even if the best flesh-and-blood financial advisors remain highly valuable.

Therefore, as a user, its crucial to take stock and be conscious that AI systems can only work from their source data combined with the information that the relevant system can reasonably request and/or receive from the user or third parties (e.g., credit rating agencies). This is the case whether the user is a sophisticated investor, or someone simply trying to make a simple but impactful decision such as picking the right pension. This consideration should be balanced against the fact that, more likely than not, there will be no meaningful legal safety net if the AI system provides inaccurate or flawed advice or services.

The Low-Hanging Fruit

Not all financial applications of AI tools are equally risky or complex. So, as a potential user, it's worth asking the question – *With respect to my own money, how can I use AI tools, right now?* To answer this, we can consider tools which are either AI-powered or rely on significant automation and are immediately available globally to most adults. As you'll see in the following sections, some AI systems or use cases for AI systems may be low-risk, while others are higher-risk but still easily accessible. These tools are often free to use and may make money by either charging to upgrade to a more "pro" version of the same tools with more functionality, or have fees of some kind for managing a service. What is important to remember is that these tools may be more easily accessible than a real person offering a similar service or function, but broadly have the same risks and are fallible.

Budgeting

Planning a budget and tracking your spending is one of the lowest-risk and easiest uses of AI tools. AI-powered budgeting apps to track your expenses automatically are readily available. In many cases, these tools are free to use, or the AI functionality is attached to existing tools.[15] Budgeting and spend-tracking tools can provide you with valuable insights and recommendations on how to optimize your spending based on patterns, statistical information, and trends.[16] They can even give you intelligently timed and targeted reminders, so you know when you have a major expense coming up and plan accordingly.

But, users should be careful to compare the options carefully and watch out for tools (whether in the form of apps or websites) that don't make assurances in the marketing or even in the fine print of the T&Cs about how they use your personal or financial data. A budgeting tool, or a tool which tracks your spending, has the potential to gather, store, and use some of your most sensitive information. Some jurisdictions have brought in regulations (notably within the EU and UK) to limit the ability of

---

[15] https://web.meetcleo.com/budget.
[16] Von Aulock, I. (2024) Top AI Budgeting Apps in 2024: Detailed features & Pricing — Invested Mom. www.investedmom.com/blog-2/top-ai-budgeting-apps.

companies to use or sell such data to "evaluate" your creditworthiness or other related sensitive assessments.[17] Nevertheless, that is the law (mostly new, mostly untested), and what happens in practice often differs.

If a tool which helps you plan or track your spending (whether AI-powered or not) shares your information with a third party such as a credit rating agency, a bank, or an insurer, this could have significant impacts on your short-term and long-term financial options. What you need to know is that the providers of these tools can't *legally* share your data with third parties unless they have your explicit consent, so just be careful to read the T&Cs. If you are at all uncertain, remember that you are often able to directly email/call the provider and ask them to clarify what they do with your data. Well worded, transparent, and easy-to-understand explanations by companies on their websites or their marketing materials are often a good sign that they understand their own legal obligations (rather than being potentially negligently unaware), and are more likely to be behaving within the law.[18]

Automated Savings and Investments

While investments can be risky, it is necessary to be conscious that a huge variety of savings and investments options can be easily accessed online. New accounts can often be opened in minutes, and in many jurisdictions, credit is easily available. Most savings and investment tools already rely on some use of AI algorithms, or significant automation, to suggest

---

[17] EU AI Act adopted by the Parliament: What's the impact for financial institutions? | Deloitte Luxembourg | News (2024). www.deloitte.com/lu/en/Industries/investment-management/perspectives/european-artificial-intelligence-act-adopted-parliament.html.

[18] Laird, J. (2023) GDPR Compliance Newsletters. www.privacypolicies.com/blog/gdpr-compliance-newsletters/.

and manage your savings and investments portfolio when that service is carried out either directly or in the background by larger financial institutions.[19]

Depending on which jurisdiction it is in which you are based, you may have access to tax-advantaged schemes such as a 401k in the United States or an ISA in the UK – each of which is often available to be managed online or via an app via a wide range of potential providers. It can be hard to pick the right provider, so try to pay attention to, and compare, their subscription or joining fees for the service, as well as any transaction fees as part of the service. Legally, most jurisdictions have regulations which require savings and investments provider – irrespective of how they are accessed (i.e., in person or digitally) – to remind you that

1. Your investments may go up as well as down, so a positive return is not guaranteed

2. Up to what level your savings or investments are insured if the provider themselves experiences financial issues

3. What the limits are on how much you can deposit/withdraw

4. All fees which may be applicable to your relationship with that provider

If a provider of financial services does not have such information clearly available to you, this should be an immediate red flag. A concern is that many AI systems, particularly those not solely aimed at providing financial advice or services, do not carry such disclaimers. So a new or

---

[19] Vipond, T. (2023) Algorithms (Algos). https://corporatefinanceinstitute.com/resources/career-map/sell-side/capital-markets/what-are-algorithms-algos/#:~:text=The%20use%20of%20sophisticated%20algorithms,of%20shares%20they%20trade%20daily.

less-sophisticated investor may be less aware of how the AI system is making decisions, as well as the direct and personal potential risks, costs, and restrictions as an individual investor.[20] Once you have compared the providers available to you in your region and set up an account with them to either save, borrow, or invest money, you may now be in a position to take advantage of AI tools they provide. Most immediately useful to ordinary users and carrying relatively limited risks, many providers have AI-powered tools to assist in setting up automated transfers to savings or investment accounts from paychecks or other income based on spending patterns, predicted income, and context such as seasonality.[21]

An advanced example of this is that you may set up an automated transfer when you receive your salary for a certain percentage/amount to go directly into a particular savings or investment account. Advanced AI, particularly when connected to your spend-tracking or budgeting tools, can adjust the size of the transfer based on what the AI believes you can afford at that time.[22]

These tools are not perfect and can only work with the information to which they have access, but can be very valuable for taking some of the complex thought (or even forgetfulness) out of the equation of financial planning. Legally, it's important to remember that there is no additional mechanism or a financial safety net which comes with these tools compared to a typical/manual financial management service or where you fully manage your savings and investments personally.

---

[20] Dunaev, M. (2024) The good, bad and ugly of using AI in financial risk management. www.wealthbriefing.com/html/article.php/ The-Good%2C-Bad-And-Ugly-Of-Using-AI-In-Financial-Risk-Management.

[21] Paycheck investments (no date). https://robinhood.com/us/en/support/ articles/paycheck-investments/.

[22] Help your money grow with automation (2024). www.fidelity.com/ learning-center/personal-finance/automate-savings.

Regarding investments specifically (leaving savings and loans aside), some of the most powerful yet simple to use AI-powered tools (which are not low-risk) allow individuals to follow what investments are being made by other people. The "other people" are often other investors on the same or similar platforms who the AI has determined have a successful track record of investment which can be replicated. But remember, even if the AI has appropriately selected current or previously successful investors, that is not to say that they will continue to be successful – so whether AI-powered or not, the selection and copying of other investors is not a guarantee of financial gain.[23] Many investment platforms of this kind offer you no financial security at all, meaning that if it all goes wrong, you could lose everything you have on that platform (or more, if you have made the investments on credit of some kind). Legally, many of these platforms are relatively transparent regarding risks, and if those risks are ones you are willing to take, then these can be "easy" (i.e., having no great requirement, active thought, or action) ways to invest money.

Credit and Fraud Monitoring

AI tools can be used to monitor your credit scores (in some jurisdictions this is more legal and/or helpful than others) and receive alerts about material changes. Many providers provide personalized advice on improving your credit score based on AI analysis. Where the law becomes complicated here is that in some jurisdictions, such as the EU, credit scoring using "simple" algorithms or formulas may currently be legal.[24] Whereas, more sophisticated or predictive AI may not. In some jurisdictions, such as the United States or even China, credit and social scoring of many kinds is far more commonly utilized and legally

---

[23] www.etoro.com/customer-service/copytrading-risks/.

[24] https://noyb.eu/en/project/credit-scoring (no date). https://noyb.eu/en/project/credit-scoring.

permissible than in the UK and EU. For large global companies who create, use, or buy/sell credit scoring this creates a complex picture when they are determining how they can operate. Despite this complexity, a good and fundamental principle which may be legally compliant and ethically reasonable is – if any credit or social scoring is being carried out with respect to an individual, that individual should be fully aware, have given consent, and be proportionate/limited to the strict legal need.

AI-powered security may also work to keep your money safe, without requiring you to do much pro-actively. Many savings, investment, banking, credit, or even insurance providers now provide AI-powered features to detect and prevent fraudulent transactions.[25] The ability of an AI system to, ideally unobtrusively, receive and produce real-time alerts for suspicious activities on your accounts can be critical to preventing fraud and safeguarding your money. This benefits the individual or business user, as well as the financial institution.[26]

Both credit score and fraud monitoring can be used by a large majority of users with relative immediate ease. Providers are typically (depending on your region) legally obliged to tell you whether using their credit score or fraud prevention services will negatively affect your credit score, or somehow restrict your access to your money.

Personalized Financial Advice

Access to AI-driven financial advice platforms that provide personalized recommendations based on your financial goals, risk tolerance, and current financial situation are already very available. Even non-specialist

---

[25] Browne, R. (2024) Mastercard jumps into generative AI race with model it says can boost fraud detection by up to 300%. www.cnbc.com/2024/02/01/mastercard-launches-gpt-like-ai-model-to-help-banks-detect-fraud.html.

[26] Krieger, F. (2024) "The pros and cons for AI in financial sector cybersecurity," Security Magazine, 24 January. www.securitymagazine.com/articles/100328-the-pros-and-cons-for-ai-in-financial-sector-cybersecurity.

AIs such as those provided by OpenAI, Microsoft, and Google are capable of producing financial advice with some prompting. While it may be generally more advisable to use a specialist tool, rather than a generalist (as would equally be the case with people rather than AI tools), it is hard to know as a user how good the advice may be. Good advice can often be a matter of opinion, so we need to consider at least the basis on which the advice is given to determine if some of the fundamentals are there. This is particularly important when considering that the law will offer little-to-no protection for any investor (whether an individual or business) who invests money on the basis of what an AI system may have answered to a prompt without the safety net of a professional financial or investment advisor operating the AI.[27]

"Good" in this case means

1. The AI understanding some or all of the personal financial situation of the user/having the capability to do so

2. The absence of, or only minimal, bias

3. Comprehension and clear communication of uncertainties in investment markets (e.g., stock exchanges, property markets, commodity and currency markets) and how these may interact with each other

The reality is, as a user, it's almost impossible to know whether an AI tool is "good" fitting the criteria set out as mentioned earlier. So, it's often advisable to

[27] Scanlon, L. (2021) "Liability for AI in financial services," Pinsent Masons, 17 March. www.pinsentmasons.com/out-law/analysis/liability-for-ai-in-financial-services.

1. Only invest minimal personal resources (money or assets) into investments recommended by AI tools. Classic advice, just like gambling, is not to put in more than you can afford to lose.

2. Seek as many independent third party reviews of any tool you are thinking of using.

3. Use AI tools for financial advice which ties to your lower risk and longer term investments such as when choosing which retirement account. This kind of retirement planning needs to factor in what may best suit your needs, based on your goals, income, and expenses – but with sufficient data and prompting is well within the capabilities of many AI tools.

4. Use a tool which is tied to a major institutional set of data. For example, and AI tool for financial advice provided by a somewhat-reputable investment bank, which is regulated and therefore under greater scrutiny may give you more comfort. It may also be lower risk by potentially having a greater cross-market understanding of economic risks and opportunities. This may mean you get quite conventional advice, from a conventional provider, but is likely to be lower risk.

Predictive Analytics for Investments or Bet-Making

Related to personalized financial advice, but distinct, is the use of AI for predicting where economic market trends will go (basically – up or down), so you can choose to invest accordingly. It may make sense to explore

AI tools that analyze market trends and provide insights into potential investment opportunities, for those people who are more comfortable taking greater control of direct financial investments. However, it's likely to be far higher risk than most uses of AI for the ordinary person looking to optimize their finances.

Despite what many may say, there is no such thing as a "sure thing" in the financial world. You will find countless articles in reputable publications and event regulators which take a very strong stance against this statement[28]: All financial investments are essentially bets, some more educated than others.[29] While some investments are less "chance-based" than outright gambling (e.g., on the outcome of a sporting event), they all have an element of risk based on somewhat unpredictable factors.

AI systems are actually very good by their very nature and design at calculating odds and predicting outcomes so long as they have the *right* data quality and quantity. Gambling companies make good use of AI and algorithms when calculating potential outcomes of events and setting their odds to make a margin against bettors. Nevertheless, it is worth noting that it is almost always high risk to rely on the advice of a single source – whether a real person or an AI providing predictions as to how financial markets will behave at any point in the future.

So, a simple and pragmatic approach for someone who insists on using AI tools for predictive analytics is to use more than one tool, or use a tool and cross-check that with the perspectives of people. Single-source reliance is often a recipe for disaster in the long run in most cases, even if there may be some short-term wins. Importantly, the law in most countries is unlikely to afford a user of AI tools for investment analytics any comfort

---

[28] Light, L. (2023). *Is Stock Investing The Same As Gambling?* [online] Forbes. Available at: www.forbes.com/sites/lawrencelight/2023/04/23/is-stock-investing-the-same-as-gambling/.

[29] FCA Insight. (2019). *Investing, insurance or gambling?* [online] Available at: www.fca.org.uk/insight/investing-insurance-or-gambling.

or remedial course of action, should the investment decision go wrong and money be lost. Unfortunately, it is a recurring theme in this chapter (and this book) that where a task such as investment is by design operated by regulated professionals, unregulated businesses and individuals who bypass those professionals are not afforded the typically attached legal and insurance protections.

Tax Planning and Optimization

Tax is a complex and high-liability issue. What this means is that it is hard to get right, and the cost of getting it wrong is potentially great. In some regions, such as North America or Northern Europe and the UK, tax authorities are highly organized and resourced. However, even in regions where tax authorities are less effective at systematically securing tax revenue, they are almost universally opaque. It's very hard in many countries to know with certainty whether an individual or business is in 100% compliance with a local tax code, if there is even the slightest of complexities to their personal or operating finances.[30] Tax laws and related guidance change regularly, sometimes multiple times per year in some countries.

The complexity is enhanced when a person has multiple tax authorities to consider (e.g., state taxes, city taxes, and national or federal taxes). What is worse is that the consequences for getting one's tax affairs wrong can be severe. Ranging from "slaps on the wrist" such as simple payment of back taxes with interest, to severe fines or even prison time.

What this means in practice is that an individual's approach to tax should generally be more cautious than with respect to their otherwise more straightforward financial planning. It's one thing to lose money on

---

[30] Dorland, A. (2024) How to keep up with tax regulatory compliance. https://tax.thomsonreuters.com/blog/keeping-up-with-tax-regulatory-compliance/.

a bad investment, its often worse to face the legal consequences of a tax dispute and risk fines or even spending time behind bars. So, the use of AI-driven tax planning needs to be approached with great caution.

However, optimistically, one could also argue that as such tools increase in sophistication, accuracy, and quality of up-to-date data, usefulness is likely to increase as risk decreases. It may even be possible for AI tools (in a similar way to financial advisory services) to identify potential tax deductions and credits, while giving recommendations as to tax optimization strategies based on the level and quality of financial data you provide to the AI.[31] However, as with financial planning, it would generally be advisable to use any digital tax advisory tools (whether powered by AI or not) and have a human tax advisor provide a second opinion, or even simply rely on a human tax advisor who is professionally regulated and capable of using AI systems to provide a better service. This could still be significantly cheaper that asking a tax advisor to do a thorough manual review of your tax affairs and provide comprehensive tax planning from scratch.[32]

Some very practical uses of AI for tax planning require little to no human action and cleverly work in the background of existing systems. For example, numerous banking apps (particularly those aimed at small business owners) or corporate credit card providers already provide intelligent automated detection and categorization of purchases to simplify and accelerate transactions and business operations. If a current

---

[31] Drenik, G. (2023) "AI-Powered tax system is creating a new paradigm. Will banks and fintechs adopt the technology to help their customers save on their tax bill?" Forbes, 27 February. www.forbes.com/sites/garydrenik/2023/02/27/ai-powered-tax-system-is-creating-a-new-paradigm-will-banks-and-fintechs-adopt-the-technology-to-help-their-customers-save-on-their-tax-bill/?sh=4e225dc06cc0.

[32] PricewaterhouseCoopers (no date) Generative AI in tax: 5 essential insights for leaders. www.pwc.com/us/en/tech-effect/ai-analytics/generative-ai-insights-for-tax-leaders.html.

provider doesn't already do this, it may be worth looking to others that sort and categorize spend and receipts in order to reduce one of the most time-intensive tasks in applying for tax deductions or rebates.

On the flip side, it's worth being aware that many tax authorities are already using automated (in some cases genuinely AI-powered) tools in order to detect suspicious, unusual, or fraudulent tax structures, filings, and non-compliance – so tread carefully. In many cases, restrictions on the use of AI for various purposes, such as profiling individuals or groups of people, may have exceptions, that is, are legal, when used by law enforcement or tax authorities. This means that a user of AI-tools looking to optimize a personal tax position is generally going to be better off in the long term being relatively conservative in approach, not necessarily asking an AI for the most creative or most immediately (potentially only short-term) tax efficient approach.

Key Considerations

1. **Verification**: Remember that AI, just like a human, can make mistakes. Even when relying on an AI to help make financial decisions, its sensible to get a second opinion even if that is from a separate digital or AI-driven service.

2. **Risk allocation**: As with any financial decision, don't put more money or assets into a single pot or investment than you can afford to lose or without a clear understanding of the specific risks. AI systems do not change the fact that you can lose your money from an unwise or unlucky investment.

3. **Privacy**: Understand that your data is valuable and that the price you may be paying, even for a "free" service, may be the ability of the service provider (e.g., the provider of the AI or the underlying financial/advisory model) to use your data or inputs.

4. **Tax**: Taxation is complex and highly variable from country to country (even city to city), so it's advisable to take specialist tax advice prior to making any investment or putting assets into a particular pot or company.

5. **Local protections**: Be aware that, in many countries, some types of savings or investments benefit from some insurance, government protection/incentives, or both. A lower-risk approach may be to focus on using AI systems to optimize these prior to making higher-risk (and potentially higher-yield) strategies which may be available to you based on AI-driven or human-driven advice.

CHAPTER 5

AI for Creative Writing

Before exploring how creatives, whether acting professionally or as amateurs, can benefit from using AI – it's worth asking the question "Will AI render creatives redundant?" The answer is a resounding no. Not maybe, but in almost every case, simply no. AI systems which aid or somewhat replace human design, writing, artistry, and creativity are only as valuable as the usefulness of their outputs. The first creative human endeavor on which AI's impact is worth considering is written language and the laws and norms which protect that form of original expression.

Scarcity, Originality, and Difficulty

As AI capabilities advance, there will almost certainly be an increasing usefulness of AI-generated or AI-inspired creative art, concepts, designs, texts – perhaps even ideas. However, over the course of 200+ years of industrialization, people have always recognized that there are inherent differences between mass-produced outputs capable of being reproduced or created with simple adjustments (prompts, in the case of AI) when compared to things which are bespoke, unique, or truly original.

A print of a Renoir doesn't hold quite the same magic, or desirability, as the canvas with authentic brushstrokes. Even in the case of photography, an artistic field where there can be significant and easy replication of a uniquely taken image, there is greater intangible value from the scarcity that comes from a limited collection of curated prints, than an unlimited printing of other images. This is even more nuanced

H. Borovick, *AI and the Law*, https://doi.org/10.1007/979-8-8688-0400-7_5

with written text where a handwritten poem by a great writer commands greater interest and value than a typed-up version. This is not just because of scarcity, but because many people recognize there is an emotional difference to the reader between a copied text and the original by the author's hand.

Although there are some robots capable of artistic and creative feats, robotics are still significantly behind the capabilities of software. Assuming this gap remains, which is a strong and likely assumption, it may be quite some time before AI systems are able to output into physical dimensions with consistent and appreciable skill, as opposed to purely digital dimensions, with anything like the level of inventiveness or precision which has already been so impressive. At the very least, purely digital creative works seem more likely to demonstrate greater inventiveness and innovation when compared to the current physical output of robotics. So, sculpture, metalwork, carpentry, drawing, painting (with real brush strokes), or other physical creative endeavors in the "real world" are less at immediate risk of genuine competition at scale from AI systems even if some technical capability exists.

Artistic design and AI are more specifically discussed in the next chapter. However, the written word is a creative pursuit as clearly as any other artistic creations. What is slightly odd is that handwritten text, the "real world" output of language, is likely to be bypassed altogether as a potential application for AI systems controlling physical objects. AI systems have been able to understand handwriting, that is, to process an image of handwritten text and convert that into a digital format which is then processed and understood.[1] Despite some robotics being technically capable, it seems redundant for a physical machine to attempt to train a

---

[1] Transkribus – Unlocking the past with AI (no date). www.transkribus.org/.

system to dexterously hold a pen and to hand-write when the output is fundamentally the same as text generated entirely within the environment of software, other than for potentially nefarious purposes such as forgery.[2]

That isn't to say that some robotics may be capable now or in the future of doing so, but the effort seems to be futile, other than as a demonstration of capability. So, the (literally) written word is likely to remain for the foreseeable future as a human-only action, so long as the skill of handwriting is not lost in favor of more efficient and shareable text, as already discussed with respect to education. In later chapters, we will also examine video, film, and still imagery in greater depth – but text best sets the scene.

Importantly, the law recognizes the differences between even a mass-produced item created by the imagination of a human being, and the mass-produced (or even limited production) output of something created by or with significant aid from an AI system. This distinction is a recurring theme throughout each chapter of this book, when considering the risks and benefits of using AI systems and the extent to which they are capable of genuine creativity in the eyes of the law.

Realistically, for the next decade, physical tasks are achievable through pre-programmed analysis and decision-making in a manner which assists or carries out tasks, rather than physical outputs of AI by design. The exception to this is likely to be in the less creative, but equally exciting, field of self-driving cars (a whole other and much more complex subject). This also means that the legal practicalities remain quite simple in the immediate term for physical outputs of AI systems; essentially the outputs are like any other manufactured product which is made with the assistance of any tools (whether sophisticated or technically complex),

---

[2] Chadwick, J. (2024) "AI can now copy your HANDWRITING - so, can you tell which of these was written by a robot?" Mail Online, 17 January. www.dailymail.co.uk/sciencetech/article-12973527/AI-copy-HANDWRITING-tell-written-robot.html.

and the primary legal frameworks focus on the originality, confidentiality, and extent of innovation in the human input. This extends to the physical printing of literary works (mostly irrespective of length). This is an important point to keep in mind when discussing the creation of text by, or with, AI systems, as it is far too easy to simply consider that output as a purely digital creation.

The Laws of Creation and Protection

If we understand that scarcity, limitation, and uniqueness are at the core of the financial, and perhaps creative, value of art – then it naturally follows that protecting those features is inherent to preserving value. The way that artists are typically able protect their art is through the typical mechanisms of intellectual property "IP" protection: copyrights, trademarks, design rights, and in some cases, patents.[3] Collectively, these are often referred to as "IP rights."

Depending on the region in which the artist/creator is creating their art, where the recipient of the creation is, and where the creation is potentially displayed or sold, the IP rights may be a mix of registered and unregistered rights. Unregistered rights simply exist by virtue of the creation being made. No further action is needed, in many cases, for those rights to exist and be exercised. Registered rights typically arise by virtue of being put on some kind of official register, and often require actual acceptance by the operators or guardians of that register. Registered rights do not arise and cannot be enforced without that registration taking place. In some countries, and for certain types of IP rights to have some validity, the registration does not need to be fully accepted but some kind of application needs to be made – you may have heard terms such as

---

[3] GOV.UK. (n.d.). *Intellectual property and your work.* [online] Available at: www.gov.uk/intellectual-property-an-overview.

"patent *pending.*" Fundamentally, if you hold one or more IP rights and accordingly benefit from the resulting protections with respect to a creative work, then you have the right to do what you want with that work within the bounds of the law. Notably, you can make money from that work, and prevent other people from doing so.[4]

The most important thing to remember is that in many regions, including the UK, the EU, and the United States, the onus is generally on the people or companies seeking to enforce their IP rights over their creation who need to prove

(a) That the rights validly exist

(b) That they own the works and the associated rights and

(c) That someone who doesn't is using all or at least a substantial part of it without sufficient permission

This means that if your creation (e.g., a poem drafted from scratch) is copied by someone else and you want to restrict them from publishing, profiting, or reproducing their version or copy of your creation – the burden of proof is likely to be on you.[5] So, having a strong understanding of this legal framework is essential if you are ever acting in a creative capacity with a real output, particularly when doing so as a professional with the intention to monetize. There is no active prevention of a third party using your creation, no person or body which is proactively seeking and clamping down on authorized use and misuse of copyright material. It's up

---

[4] Dennis, G. (2024) "UK copyright law: the basics," Pinsent Masons, 9 January. www.pinsentmasons.com/out-law/guides/copyright-law-the-basics.

[5] Intellectual Property Office (2014). *Enforcing your copyright.* [online] GOV.UK. Available at: www.gov.uk/guidance/enforcing-your-copyright.

to you to find out that there is potential misuse of your creation (i.e., infringement of your IP rights), somehow prove that you own the IP rights to your creation, then take potential legal or business action against the infringer.[6]

If you are successful in taking a legal action, depending on your country of residence and the country where the misuse took place, you may be entitled to one or more legal remedies such as financial compensation. Most importantly, if successful in a legal action, you may be able to simply ask the courts to prevent the misuser from making further use of your copyrighted work. It is worth bearing in mind that different countries not only have different rules regarding IP rights but, to benefit from protections outside of your own country, you may need to make legal IP registrations for IP rights such as copyright or trademarks in multiple territories.[7] This can be expensive. In some cases, there are international agreements which may give you automatic rights in multiple countries, by virtue of original rights in one (likely where you live or where you made the creative work).

While there is a financial cost to register and benefit from many types of IP rights, particularly where international protection is sought, it does mean that if you wish to use the law to your advantage to protect something original you have made, many options are available to you. There are sufficient mechanisms to allow you to decide if you only want to rely on free or cheap IP rights or consider registration of additional protections domestically and internationally.

---

[6] CopyrightUser. (2017). *Someone Infringed my Copyright.* [online] Available at: www.copyrightuser.org/faqs/question-9/.

[7] U.S. Copyright Office (2019). *Copyright in General (FAQ) | U.S. Copyright Office.* [online] Copyright.gov. Available at: www.copyright.gov/help/faq/faq-general.html.

It's not just the registration processes of IP rights which may differ internationally. The very nature of what IP rights actually are, and how they arise, varies from region to region.[8] Appreciating that the differences and nuances can be complex, this is a good point to say – if in doubt, seek the advice of a specialist lawyer in this field (such as a patent attorney, trademark attorney, or other similar IP legal advisors). At the very least, it is worth doing your own research, potentially with the help of an AI system, cross-checked with other sources, to understand whether you need/ want to take any actions to protect your creative works. For example, in the United States, creatives benefit from image rights which don't exist to quite the same extent in the UK or EU.[9] In the UK, the subject of an image doesn't necessarily own that photo/drawing/recording – it's usually the creator. This is often controversial in the case of a photographer who might own a lucrative or iconic photograph of a person as the photographer may financially benefit from the image of another person without the subject of that image having any rights to share in the profits.

Case Study – Rihanna v Topshop

The complexities of international unregistered and registered IP became a particularly notable issue in the famous case of Rihanna v Topshop, where the famous singer (and now tycoon) brought a claim in the UK against a major clothing retailer.[10] The retailer licensed an image of her from a photographer who owned the photo. The default position is that a photographer owns the image they capture or create. This applies to the written word as well – if you write something and you are not contractually

---

[8] Park, S. and WIPO (no date) *Intellectual property issues in international business*, *WIPO*. report, pp. 1–25. www.wipo.int/wipogold/en/.

[9] U.S. Copyright Office (2024). *What Is Copyright?* [online] www.copyright.gov. Available at: www.copyright.gov/what-is-copyright/.

[10] Pop star Rihanna wins image battle. (2015). *BBC News*. [online] 22 Jan. Available at: www.bbc.co.uk/news/entertainment-arts-30932158.

bound to give ownership of that work to someone else, it is yours. In this case, Rihanna had to rely on a different mechanism, which essentially only exists in the UK called "passing off."[11] This meant that Rihanna and her legal team, at great cost, had to sufficiently demonstrate customers seeing the t-shirt in a shop would be misled to presume she had a brand deal or endorsement with the retailer. Clearly, this is a complex and particular case and most people protecting their IP rights, whether in their own image or in an original piece of literature, will have far fewer resources to spend on enforcing their rights (which are likely to be simpler).

When it comes to AI, IP rights are more complex. This is mostly due to the uncertainty in essentially all legal systems as to how automatously creative and capable of genuine inventiveness AI systems actually are. Of course, as the capability of AI systems grows, perspectives on this may also change. There are two key questions which complicate the quagmire of existing IP protection for creative works.

The Two Big Questions

Question 1

Are AI-generated* outputs capable of, or have the right to, benefit from any kind of IP protections?

Question 2

Is the use of AI systems to make a new creative work potentially in breach of other people's IP rights**, if the underlying data within those AI systems contains creations and works of others?

Generation ranging from the merest sniff of AI in the process to being fully produced by AI

**By the user of the AI system and/or the creator of the AI system*

---

[11] Cran, D. and Mark, B. (2015) *IP Alert: Court of Appeal confirms Rihanna's image protected under the "umbrella" of passing off.* www.rpc.co.uk/perspectives/ip/ip-alert-court-of-appeal-confirms-rihannas-image-protected/.

The strict tests for whether AI-generated creative works are capable of benefitting from IP protections varies from country to country. A good starting position, and general rule of thumb, is to question whether the creative works have "sufficient human authorship."[12] What is "sufficient" is a question within a question, that the legal systems of most countries have not quite figured out and settled.

There is a possibility that IP rights may be valid and enforceable if an AI system (or a tool which is somewhat powered by AI) provides assistance with limited tasks. This may be the case despite an AI system clearly not being a "human author." For example, with respect to creative writing, the organizational planning of a chapter of a book where the contents are already fully fleshed-out, but where the AI system does not create or generate any of the content. In some countries, it may be possible that further assistance such as limited drafting or re-drafting may not invalidate the IP rights in those creative works, and therefore the ability to protect them from misuse by a third party. However, this is not a hard-and-fast rule. It is quite possible that, depending on region/country, the slightest whiff of an AI system interacting or impacting on the creative process could be sufficient to compromise the ability of a writer to preserve IP rights.[13]

Really, the only near-certainty with respect to AI systems and IP rights is that AI systems cannot be considered creators in their own rights. This means that something created by an AI system with little-to-no human

---

[12] Oratz, L. T, Hameen, J. and West, D. S. (no date) *Human authorship requirement continues to pose difficulties for AI-Generated works.* www.perkinscoie.com/en/news-insights/human-authorship-requirement-continues-to-pose-difficulties-for-ai-generated-works.html. (Accessed: June 9, 2024)

[13] Rosati, E. (2022). *US Copyright Office refuses to register AI-generated work, finding that "human authorship is a prerequisite to copyright protection."* [online] The IPKat. Available at: https://ipkitten.blogspot.com/2022/02/us-copyright-office-refuses-to-register.html [Accessed 10 Jun. 2024].

input or guidance is not an original work capable of protection. AI systems are generally perceived by regulators, IP registers, and courts as tools rather than autonomous entities – for now.[14]

Courts in the United States have been notably hesitant to recognize the validity of IP rights when there has been significant AI involvement in the creative process. This is the case despite US IP registers, such as the USPTO, providing guidance which is increasingly relaxed with respect to the involvement of AI systems in the creative process as long as there is significant human contribution.[15] This makes for a confusing picture, and means there is legal uncertainty as to whether some IP rights are treated differently from others, even with the same countries, depending on the guidance of the relevant register and the determination of a particular court. By contrast, the UK courts have seemed more consistent, and somewhat lenient. However, an IP right you can only enforce in the UK, when the United States may be the location of the infringement by a third party, isn't always particularly helpful.

Even if a writer is trying to enforce IP rights over their creation in a more AI-friendly jurisdiction, there are significant practical problems. These problems are notably increased in less AI-friendly IP jurisdictions because it may be very hard to prove or trace from an evidential perspective that an AI system was used to an extent that there was no significant human contribution. Short of trawling through the creator's Internet records and AI system logs, which may also be inconclusive, there is an element of evidentiary uncertainty for either side of a legal dispute which is difficult to resolve. As legal disputes over IP rights in

---

[14] Harris, B.J.C., Lindsay (2024). *UPSTO Issues "Significant" Guidance on Patentability of AI-Assisted Inventions, but unlike USCO, Does Not Require Disclosure of AI Involvement.* [online] Cleary IP and Technology Insights. Available at: www.clearyiptechinsights.com/2024/02/upsto-issues-significant-guidance-on-patentability-of-ai-assisted-inventions-but-unlike-usco-does-not-require-disclosure-of-ai-involvement/ [Accessed 10 Jun. 2024].
[15] Kim, C. *et al.* (2024) *Inventorship guidance for AI-assisted inventions.* USPTO.

creative works are brought by the claimant, that is, by the party who claims they own the work and is seeking legal protection, an inability to prove that AI did not excessively diminish the human authorship can be very challenging. Even if there is evidence that an AI system provided suggestions as to organization or structure of a book, article, or story, it may be very hard to demonstrate that such suggestions were actually applied, or that the final output by the writer materially reflects the recommendations of the relevant AI systems.

So, it may be sensible to look at the considerations and potential answers to the key questions within the specific framework of each creative undertaking (i.e., writing being distinct from drawing – the latter being considered in the next chapter). We can form some general conclusions based on the state of the laws in many regions/countries which have experienced significant legal explorations of these questions applied to AI:

1. Are AI-generated outputs capable of, or have the right to, benefit from any kind of IP protections?

    a. **Potentially yes in many (but not all) countries if an AI system was used to a limited extent as a tool at some point in the creative process and did not copy someone else's original works.**

    b. **Possibly, if an AI system or tool was used to facilitate an initially loose concept, plan, or design – but not actually create the drafts or final version, or if those drafts and final versions still involved significant human contribution so as to be considered "sufficient human authorship."**

    c. **Almost certainly not if the outputs were materially conceived, designed, or created by an AI system or tool using a human-made prompt with limited further human contribution, or if the AI outputs copy other original works.**

2. Is the use of AI systems to make a new creative work potentially in breach* of other people's IP rights, if the underlying data within those AI systems contains creations and works of others?

    a. **In most cases, using many currently available tools which draw data from billions of sources – possibly yes, there could be a breach.**

    b. **Where there is a clearly drawn influence or direct copy from existing works – probably yes, there is likely to be a breach.**

*A misuse of the underlying creative works which may be contained in the AI system's data by the user of the AI system and/or the provider of the AI system

So, with these questions in mind, it is important to try to practically consider the day-to-day uses of AI systems for creative writing, particularly where the creative works are intended to be published.

Writers and Journalists

The AI system which is most credited as bringing AI into the public consciousness as a daily tool is undoubtedly ChatGPT. The launch of GPT-3 was groundbreaking. On the other hand, the breakneck pace of change in AI is exemplified by how quickly and resolutely GPT-3 has since been surpassed in terms of ability, precision, and complex task handling (even by the immediate successors produced by its owner, OpenAI). ChatGPT and other similar systems are primarily text-in = text-out systems. This has significantly advanced since its release, with newer AI systems including capability of text to image, as well as text-in or audio-in to any permutation of text, image, video, and audio response.[16] For the purposes of this chapter, focusing on text outputs, writers and journalists can generate significant volumes of text almost instantly. The quality and accuracy of that text output will vary significantly based on where the AI system provider sources its underlying dataset, and how the AI system's core models are trained.

With the example of ChatGPT and similar **mixed input-in = text-out** AI systems in mind, we can consider practical expansions of the big two questions we have already considered:

1. *Where do the text outputs come from; are they actually original or a mish-mash of mixed sources?*

2. *Is the text drawn from an understanding of one or more sources which are other people's original works and writings?*

3. *Who owns the text that is produced?*

4. *Can a user do anything they want with the text that is produced (i.e., does the user own the unregistered IP rights, and could they register for IP rights in the outputs as creative works)?*

5. *Is a something ethically or legally wrong by using the AI system to generate the outputs?*

Thinking through these questions is easiest when applied to a practical (hypothetical) example scenario.

---

[16] Kerner, S.M. (2024) GPT-4o explained: Everything you need to know. www.techtarget.com/whatis/feature/GPT-4o-explained-Everything-you-need-to-know.

Example Scenario

A writer is thinking about using an AI system from a provider such as OpenAI, Google, or Anthropic. They do not check the terms and conditions of the AI system provider to understand from where the underlying data (or consents and licenses for that data) in the AI system has been obtained. They also do not check what happens with the data they may put into the AI system as part of a prompt.

STAGE 1 – A writer has an idea for an article and jots down some innovative, original, but disordered bullet points to outline their idea. Some of the research relevant to the article is likely to include sensitive, confidential, or personal information.

STAGE 2 – The writer then asks an AI system (for ease, lets imagine it is a text-in to text-out AI system such as ChatGPT4) to help organize the bullet points for an article.

STAGE 3 – The writer then asks the AI system to re-organize the bullet points again to create the structure for the purposes of a 2000-word article in a style which would be aimed at a particular category of people (for example, a hypothetical audience with a short attention span, with an average reading age of 11).

STAGE 4 – The writer drafts a good, if oversized, 2500-word article and asks the AI system to slim it down without losing the key points to meet the 2000-word count limit.

STAGE 5 – The writer decides they are unhappy with the final article. So, they decide to draft it from scratch. This time, entirely written by an AI system after giving it a series of detailed and specific prompts, containing extensive information.

Considerations Applied to Scenario

Although the law is quite unsettled in most jurisdictions, and likely to experience lots of challenges and divergences based on regions/countries, from this scenario, we can make some reasonable assumptions and predictions (with confidence but not certainty) that

A. **If the journalist had stopped at STAGE 1** or written a typically researched article without further assistance from AI systems

- The rights to the article would have been theirs, unless they had an agreement with a third party (presumably, a publisher) that they were writing it on assignment for them to own.

- The article is likely to be sufficiently original to be copyrightable and to be at low risk of infringing the copyrights of others. Therefore, with relative confidence, it can be used, published, licensed, or sold.

- Is unlikely to expose the journalist to ethical or legal risk (unless the contents are themselves unethical or potentially illegal – but that's not the fault of an AI).

B. **If the journalist had stopped at STAGE 2** and written a typically researched article without assistance, beyond the organization of bullet points, from AI systems

- All of the points set out at the preceding A likely still apply.

- It's unlikely that someone would seriously believe it is unethical to ask an AI system to re-organize existing/originally drafted language.

- Irrespective of a lack of ethical risk, there is a potential legal risk that if the writer ever wanted to enforce a copyright (i.e., in a court against an infringer/copycat), they would have to prove the originality of the article.

- The involvement of an AI at any stage, no matter how limited, may have some impact on the ability to enforce the copyright. Depending on which region/country it is in which the writer is trying to enforce their copyright, the AI in the mix could be material or trivial to the decision of a court.

- The journalist may have also breached confidentiality or data privacy obligations to any relevant third parties whose confidential, personal, or sensitive information were included within the bullet points.

C. **If the journalist had stopped at STAGE 3** and used the AI system to create a structure on the basis of specified criteria, but then proceeded to write the substantial majority of the article themselves

- As we already know that the involvement of AI at any stage is a potential risk, irrespective of whether the AI system generated any new content/material as part of the re-structure – the risk has increased compared to STAGE 2.

- Depending on the jurisdiction in which the journalist publishes the article, it may be hard for the journalist or publisher to enforce a copyright should someone misuse the article.

- It is arguable a *substantial* creative element of the article (i.e., the plan/structure) was AI-generated, even if this is not the case in terms of the actual number of words on the page. Therefore, there is a question about whether there is, by contrast, sufficiently substantial human authorship.

- In some jurisdictions, this may still be sufficient to secure a copyright or some IP rights, but will vary.

- The journalist may have also breached confidentiality or data privacy obligations to any relevant third parties whose confidential, personal, or sensitive information were included within the bullet points (as with STAGE 2).

D. **At STAGE 4**, the journalist has essentially handed at least a substantial portion of the creative control and output to the AI system.

- This delegation of editorial control seems at least substantial, arguably almost total.

- The journalist, though still the source of the original idea, has likely given up their ability to enforce IP rights regarding the article (or even snippets of its contents) in many jurisdictions.

- They may still benefit from some rights or protections in certain jurisdictions, but their ability to globally publish and monetize the article will be subject to a materially greater risk of preventing others from copying/using it.

- The journalist may have also breached confidentiality or data privacy obligations to any relevant third parties whose confidential, personal, or sensitive information were included within the bullet points (as with STAGES 2 and 3).

E. **At STAGE 5**, the journalist has granted almost the entirety of the "pen to paper" aspect of the creative process to the AI system having only acted creatively and with originality with respect to the prompts.

- Initial editorial control is entirely delegated to the AI system.

- Even if the journalist heavily edits the article, the substantial core of the article was produced by the AI system.

- Similarly to STAGE 4, the journalist may still benefit from some rights or protections in certain jurisdictions but they are at significantly greater risk of being unable to enforce IP rights against any third parties or have an exclusive ability to monetarily benefit from the article.

- The journalist is at potentially serious risk of having infringed the IP rights of other writers, whose original creative works (into which the IP rights apply) may have been misappropriated or misused by the AI system provider without the other writers' permission.

- The journalist may have also breached confidentiality or data privacy obligations to any relevant third parties whose confidential, personal, or sensitive information have been included within the bullet points. This is similar to STAGES 2, 3, and 4, but the severity and impact of the breach could be greater if the prompts to the AI system contained more confidential, personal, or sensitive data (by volume or qualitatively) than in the original bullet points.

While there is an almost continuous stream of new legal, regulatory, and industry guidance (as well as actual laws being drafted, considered, and in some cases, passed), it's very hard to suggest with certainty on whether a writer can protect their output when using AI. Some more recent positive regulatory changes indicate that the tide is turning. Being somewhat optimistic, before too long the use of AI will be considered as ordinary as the use of a laptop instead of a pen, and therefore equally uncontroversial to the protection related to IP rights. The law will catch up to some extent, but it is an unfortunate reality that it always lags behind technological progress.

Realistically, it is quite likely that a piece of creative writing which arises, as described in the preceding STAGE 5, will not be protectable as a piece of IP in the medium or even long term unless there is a significant change to how we conceptually consider IP itself, that is, diverging from the idea that it is tied to human original creativity. Nevertheless, in the medium to long term, various regional and international legal and regulatory trends indicate that creative writing, which isn't entirely, or nearly entirely, created by AI, will benefit from the same IP rights automatically arising or available for registration in entirely human-generated writing. While it is probable this will happen at some point, most likely incrementally, it would be risky to expect those IP rights to be enforceable with certainty in the immediate term.

One of the biggest reasons for the uncertainty as to how IP rights can arise and be enforced in AI-generated outputs, particularly written creative works, is due to the quagmire of complex issues around where the data is sourced that trained the model that writes the output, and whether it is possible for AI systems to truly create something without "copying" (for lack of a better term) the creations of another human author.[17]

Infringement of the Rights of Others

It is worth noting that in the 5-Stage example of the creative writing process set out earlier in this chapter, there are additional risks beyond the inability to enforce one's own creative IP rights against others who might copy creative work or aspects of its contents. The consequences of these risks may actually be greater to users of AI systems than the potential inability to protect their own creative works. Use without permission, whether intentionally or unknowingly, of other people's creative works may directly expose users of AI systems to legal liability. This could be direct liability, irrespective of whether the AI system provider can be proven to have acted in breach of people's IP rights. As we know that AI systems essentially work on the basis of their source data, greater involvement or demand on the AI system to create the actual text is likely to lead to a greater probability of infringing the rights of others.

---

[17] Trigg, R. *et al.* (2023) *Generative AI: how sourcing data for training AI tests UK and EU intellectual property rules | Osborne Clarke.* www.osborneclarke.com/insights/generative-ai-how-sourcing-data-training-ai-tests-uk-and-eu-intellectual-property-rules.

Why is this a risk? Many of the texts that AI systems have collected or used have been collected legitimately in large data-gathering commercial acquisitions or with contractually purchased licenses.[18] However, there are an increasing number of legal claims by publishers, creative writers, artists (and basically anyone who may think they've created something original on the Internet) against some of the companies that own and operate AI system. The basis of these legal claims is that large tech companies, including those building, training, and providing AI systems, have illegitimately acquired data by scraping the Internet and various digital sources without the consent or permission of the content owners (i.e., those who hold the relevant IP rights in creative works).

Although lots of the data on the Internet is not proprietary (meaning legally owned by someone), there are billions of creative works which are online and may benefit from IP rights which are being used by AI providers without total transparency and consent. This also means that the owners of those creative works are generally not seeing any financial benefit from their IP rights, a key reason IP rights attach, or can be attached, to creative works. So, they are incentivized to take actions against AI providers, but they are not limited from taking action against users who they can prove have copied their works or created clearly derivative works without consent and/or payment to the original creators.

---

[18] Gulp Data (2023) The asset behind the AI: How data plays into OpenAI's $10B payday from Microsoft. www.linkedin.com/pulse/ asset-behind-ai-how-data-plays-openais-10b-payday-from-microsoft-/.

Case Study – The New York Times v OpenAI

The case of The New York Times (NYT) v OpenAI[19] is a clear example of this creator vs AI provider tension, and is notable because of the NYT's arguments that not only has OpenAI infringed their copyrights – but that by doing so, it actually caused dilution in the value of the NYT's brand while creating unfair competition.

Essentially, NYT was directly accusing OpenAI of giving away the NYT's content (and that of its writers) for free or cheaply, while the NYT spent the overheads to create the content and would charge its readers for it. Essentially, misappropriation and misuse of the NYT's (and it's writers') creative works, therefore allegedly breaching their IP rights over those works. This case, and similar cases of its kind, have two sides (like any legal claim). One side believes that this kind of liberalization of the Internet is progressive and creates creator democratization of content creation, the other believes that content creation should and does have a cost which must be fairly compensated or quality journalism and writing of all kinds may disappear as a commercial endeavor. If we take cases like that of the NYT's to their most extreme conclusion, we could end up in a societal position where it would be no longer financially viable for investigative journalism or creative writing of any kind to be a profession as their margins are squashed by IP infringement (i.e., AI systems giving their content away for free, while also generating an automated competitor in creative writing). This would also mean that the organizations behind writers such as newspapers, book publishers, and even many websites, could face financial failure. Should there not be rapid legal certainty, either

---

[19] Grynbaum, M.M. and Mac, R. (2023). The Times Sues OpenAI and Microsoft Over A.I. Use of Copyrighted Work. *The New York Times.* [online] 27 Dec. Available at: www.nytimes.com/2023/12/27/business/media/new-york-times-open-ai-microsoft-lawsuit.html.

via resolutions to cases such as NYT v OpenAI or the passing of new laws, there would almost inevitably be an impact on the quality and quantity of content available to readers.

Accordingly, there is a real potential societal impact beyond just the debate over the protection of IP rights in the creative works. It is clear that the practical impacts of this current legal uncertainty are wide-reaching and difficult to quantify. Other than the financial and content-related impacts on society that cases like the NYT v OpenAI might have, there are other reasons that these should matter to you. Ethically, many writers and readers engage in these activities driven either by recreational enjoyment (e.g., reading a fictional book) or education (e.g., reading non-fiction books, newspapers, and websites). The incentives to create works of significant originality, or works with educational importance, significantly diminish when their content is misappropriated and distributed by AI systems without any major consequences. At the same time, technology companies are typically driven by goals of growth and revenue.

Should checks and balances against large AI providers not work effectively (e.g., lawsuits by third parties and enforcement actions by regulators), this would be a major example of unchecked power by technology companies, which is unlikely to benefit consumers in the long-term. It is therefore quite likely that the next decade will see an increasing amount of involvement and scrutiny by competition regulators (called "anti-trust" in the United States) in various countries. The impact of such scrutiny is difficult to predict, but hopefully means that there will be more practical choices for people using AI systems and less consolidation. As a result, the benefits increase of reading terms and conditions of different AI system providers before selecting one to use. If there are significant and meaningful differences between the approach and setup of different AI systems providers as a result of regulatory or commercial drivers, understanding where data comes from and where it goes will become more important, and potential users should have greater choice to suit their needs. For example, allowing a user to choose a provider who makes legal assurances that all of their source data has been legally licensed from knowing third parties – even if that technology is potentially not quite as powerful, but offering far lower risk.

In summary, despite significant legal, regulatory, and commercial uncertainty as well as great divergence of approach between countries, it is broadly clear that

- The risks are materially higher for AI-assisted writing compared to writing an original piece of content "the old-fashioned way" (however antiquated this notion may be) or using an AI system only for the purposes of helping you research, organize, and structure your own original ideas.

- The ethical and legal risks around plagiarism are more or less equivalent to the risks and potential consequences of non-AI assisted plagiarism (basically, stealing is stealing and copying is copying – irrespective of methodology or tools used). Using someone else's work without their permission, or at least crediting them, is almost always a bad idea).

- Even if legal claims against AI system-providers fail, your own risk of infringing someone remains – whether legally, ethically, or both. This risk will likely change and probably increase as new laws and regulations arise to set the guardrails, though navigating those risks is likely to become easier and more transparent.

Key Considerations

1. There are plenty of ways that AI can be used safely, legally, and in a relatively low-risk and low-cost manner to accelerate the processes of creative writing and research. There is likely to be a correlation between a writer not feeling they are *reliant* on an AI system to produce a creative work (i.e., they are putting in significant human effort), and a lower risk of legal issues for that writer.

2. The world is likely to rapidly progress such that many, if not most, writers will be using AI systems to support the production of new creative works. This likely means that many writers will face increasing pressure, by their audience and potentially by publishers, to keep up.

3. When using an AI system to generate text, users don't know where that text came from. As a writer, that has inherent risks which cannot be entirely avoided.

4. As an end user of AI systems in creative writing, you cannot be fully certain that your text is fully without content that infringes the legitimate IP rights of other writers or publishers and therefore that you will not face some kind of potential legal claim for IP infringement.

5. In the long run, if court cases start to rule against the providers of the AI systems used by writers (for example, if you used OpenAI's ChatGPT to help write an article or book), it is conceivable that the writer may be somehow legally accountable, even if the AI system provider is not.

6. The potential legal claims you could face may arise irrespective of whether a legal case arises or is won against an AI system-provider (i.e., a writer would still be copying , it doesn't necessarily matter how) At the very least, there is a significant ethical question which writers using AI systems should ask themselves: Am I doing something wrong by *potentially* copying aspects such as phrases, sentences, paragraphs, themes, or previously original ideas from someone else?

CHAPTER 6

AI and Design

Having considered the issues and impacts of AI in the creative writing process in the previous chapter, it is worth considering a similar but distinct review of the legal risks of AI when applied to design and visual creativity. For the purposes of this chapter, "design" means the process of conceptualizing, outlining, creating, using, or monetizing a creation in a visual rather than written medium. This could be a static design such as a drawing or still image, an animation, an audiobook, podcast, song, or even a full video – all of which can already be partially or fully generated using AI systems. This chapter primarily focuses on the visual medium of design, but most of the concepts and risks discussed apply similarly to design in an audio format. Some of these systems are free to use and are quite simple with limited functionalities, but more advanced design tools are often integrated or fully built on top of advanced AI systems. So, design should simply be taken to mean any kind of visual design, even if not specifically discussed within this chapter.

Originality

Working within the established premise that a mass-produced print of even the most valuable original doesn't hold the same qualities of desirability or value, it is easy to immediately draw out an immediate question a consumer of an AI-generated or assisted design might ask – "What exactly am I viewing/buying?" and "Does this have a real value?"

© Harry Borovick 2024
H. Borovick, *AI and the Law*, https://doi.org/10.1007/979-8-8688-0400-7_6

The starting point, both at law and in practical reality, is that something new has been created. An AI system has simply been used to arrive at a real piece of visual or audio design which did not previously exist in its entirety. The exception to this is where a user is actively using an AI system (and prompts it accordingly) to knowingly create an exact replica of a piece of existing design. Where the law and reality diverge is whether the new piece of design is genuinely novel and therefore whether it is capable of being protected under the law in the same way as truly unique and original design. Basically, the law is intended in most jurisdictions to preserve and reward ingenuity and originality. Similarly, in practice, it is typical that design which is original and innovative is more valuable, particularly when it is limited or can be protected and then sold in greater numbers (e.g., an original song which can then be sold/streamed).

Recall that the typical setup under the laws of most commercially minded jurisdictions, that is, where there is a strong rule of law to protect business growth and freedom of expression, is that the people or companies seeking to enforce their IP rights over their creation need to prove:

(a) That the rights validly exist

(b) That they own the works and the associated rights and

(c) That someone who doesn't is using all or at least a substantial part of it without sufficient permission

Rights That May Exist in Designs

IP rights that may be available or automatically arise in designs are often wider than for text-based creative works. While there is international variation, there are numerous types of rights (i.e., (a) in the preceding list) that may be available or even overlap.

The main types of rights over original designs are

1. Copyright (similar to text-based creative works)

2. Design rights (varies significantly between jurisdictions)

3. Trademarks (similar to text-based creative works)

4. Patents (varies significantly between jurisdictions)

5. Moral rights (varies significantly between jurisdictions)

Some of these rights are commonly sold as an asset in and of themselves. For example, the inventor of a creation who secures a patent may sell the rights in that patented invention (i.e., the right to benefit from the original idea) even if they are not selling the actual manufactured end product.

While "design rights" and "moral rights" don't even exist in many jurisdictions, it is possible to have some form of IP protection in most countries via one or more of copyrights, trademarks, or patents. It is important to distinguish the confusing terminology of "design rights" (which can have a strict legal meaning, as one type of IP right) and IP rights generally (such as copyright) which may apply to original designs.

Taking these individually:

1. **Copyright – Common, globally.**

   Copyright is the most well-known IP right applicable to creative designs. Copyright protections don't act to protect ideas, those are much harder to prove and protect. Rather, copyright covers the output from an idea (i.e., the tangible creation only).

Copyright is a powerful right. In many jurisdictions,
the right lasts for the length of the creator's lifetime
+ some additional years (70 years in the UK). It
is worth noting that the duration of copyright
protection differs in creative visual (or audio) pieces
vs. written works.[1]

Like most IP rights, copyright is assignable. This means that the creator has the ability to not only sell copies or pieces stemming from the original copyrighted creative work but also benefits from the ability to sell or transfer the actual core copyright in the original. Once that core right is assigned (e.g., sold or otherwise given away) the original creator generally won't be entitled to any further benefit.

Copyright and AI

With respect to AI-generated creations, it is seriously unsettled whether a copyright can exist in an "original" piece created by AI irrespective of the level of human creativity also involved. Crucially, copyright protection is typically recognized by many legal systems to only apply where there is sufficient (or even, in some cases, total) human authorship.[2] As copyright doesn't protect the idea, only the output – it also doesn't help the creator protect the human element, that is, the idea, if the creative work is substantially AI-generated.

---

[1] Government Digital Service (2014). *How copyright protects your work.* [online] GOV.UK. Available at: www.gov.uk/copyright/how-long-copyright-lasts.
[2] www.jonesday.com. (2023). *Court Finds AI-Generated Work Is Not Copyrightable.* [online] Available at: www.jonesday.com/en/insights/2023/08/court-finds-aigenerated-work-not-copyrightable-for-failure-to-meet-human-authorship-requirementbut-questions-remain.

2. **Design rights – Uncommon globally, but commonly relied-on in the UK and EU.**

> Design rights are often used to protect the appearance, structure of, or aesthetic of something functional. Design rights focus on the unique aspects of shape, texture, ornamentation, or configuration of a visual design.[3] Like copyright, depending on the jurisdiction in which the creator operates, or where a design is first made public, they may be able to rely on either registered or unregistered design rights in addition to copyright protections. Design rights are capable of being less-specific than a copyright.

So, whereas a copyright may cover a whole design and the specific details of it, design rights may mean that a separate or alterative coverage is also available – such as just the outline or shape of a design. Whether a design right or copyright is available or appropriate as an IP protection is usually quite fact-specific. On the other hand, design rights are also more limited. Design rights tend to protect three-dimensional shapes and configuration of objects (depending on the relevant country), whereas copyright tends to protect two-dimensional artwork and literary works.[4] They do not apply to many forms of visual or audio creative works in the same way as copyright. So, videos, photographs, and music cannot benefit from design rights.

---

[3] www.ouryclark.com. (n.d.). *Design Rights.* [online] Available at: www.ouryclark.com/resource-library/quick-guides/intellectual-property/design-rights.html#:~:text=Design%20rights%20protect%20the%20appearance.

[4] Steele, C. (2020) *Is copyright riding to the rescue of designers?* www.ashfords.co.uk/insights/articles/is-copyright-riding-to-the-rescue-of-designers.

If a creation does not have a functional element, that is, if the creation is purely artistic, it generally will not qualify for protection as a design right. Naturally therefore, some of the most common people or companies who seek to rely on design rights are inventors, due to the requirement of a functional element to a design. Typically, design rights have a shorter period of protection than copyright. For example, "unregistered community design rights" in the EU only last for three years after the creation of the relevant piece.[5] In the United States, or in any jurisdiction where design rights only have relatedly short-term protection, it would be most typical to rely on some kind of design patent (a much stronger right which is more complex to register), which can give significantly longer protection.

Design Rights and AI

You may wonder why such a similar right exists alongside copyright. Simply, design rights allow for some level of formal registration which copyright in many countries doesn't, typically where the design is primarily functional and three dimensional. Whether registered or unregistered, the requirements of design rights typically mean they apply to products rather than media or artistic creations. Accordingly, registered design rights should in many circumstances make it easier to protect a design via legal enforcement actions and therefore monetize it and consider the design as an "asset."[6] If you're a creator, looking to use your designs on a professional basis (i.e., you want to sell things, or build a business with clear ownership of designs), this could be incredibly important. Crucially, design rights and

---

[5] www.ouryclark.com/resource-library/quick-guides/intellectual-property/design-rights.html#:~:text=Design%20rights%20protect%20the%20appearance,ornamentation%20of%20the%20product%20itself.

[6] Prior, F. (2022). *Copyright or Design Rights? | LegalVision UK*. [online] legalvision.co.uk. Available at: https://legalvision.co.uk/intellectual-property/design-rights-vs-copyright/#:~:text=Design%20rights%20and%20copyright%20are [Accessed 16 Jun. 2024].

copyright overlap – but don't exclude each other. So, a creator is not likely to compromise one by seeking the other.

Accordingly, if a user of AI tools generates a design and there is uncertainty as to whether reliance can be placed on a copyright (particularly if it has not yet been challenged or needed to be enforced), then attempting to rely on a design right registration may give the creator a strong indication as to whether the design is capable of being protected at all. If so, then the creator will benefit from a shorter-lived baseline of protection.

A point of legal technicality worth noting is while an AI system itself cannot be the legal owner or beneficiary of design rights in an AI-generated design (due to a lack of legal "personality"[7]), a creator who uses AI to create a design may be able to register a design they have created using AI. The reason this is currently a minor point in the law is that practically most creators do not want an AI system to take the credit/benefit for an AI-generated design instead of the person using the system.

3. **Trademarks – Common, globally.**

    Trademarks are arguably the simplest registerable IP rights for a visual design which carry significant advantages. Trademarks, unlike copyright and some types of design rights, require registration. The reason they are simple is that they don't always require real innovation or creativity to be capable of registration. However, they do need to be registered. So, no registration means no protection.[8]

---

[7] Fotheringham, C. and Guild, J. (2023) *AI and design rights – ownership queries.* www.ashfords.co.uk/insights/articles/ai-and-design-rights-ownership-queries#:~:text=As%20AI%20does%20not%20have,or%20owner%20of%20a%20design.

[8] Mewburn (n.d.). *UK Trade Marks - The Basics.* [online] www.mewburn.com. Available at: https://mewburn.com/law-practice-library/uk-trade-marks-the-basics. [Accessed 16 Jun. 2024]

Trademarks are designed solely to protect the individual identifiability of a brand, product, service or business. They are often used to protect brand identifiers like logos and names, without needing to have true originality or creativity in their creation process. They simply need to be sufficiently distinct and used/usable in a manner which is clearly connected to the person or business seeking to register and enforce them. So, because trademarks lack the same kind of artistic or creative foundations as many other IP rights and they simply need to be sufficiently unique to the relevant service or industry to which they attach, they potentially lend themselves very well to use in and around AI (or even creation by AI). Trademarks can, in many countries, be registered and maintained indefinitely as long as they are renewed.

Trademarks and AI

Clearly, the long-term or indefinite shelf life of a trademark as an IP right makes trademarks particularly valuable as assets with a high potential reward when balanced against the potential risks of applying of AI to designs. A logo created using AI tools, or which is fully AI-generated, may, in some cases, be capable of being trademarked – even if it is not copyrightable or capable of benefitting from any other IP rights. This does mean that it is imperfect with narrow applications and benefits. However, it is quite likely that in the years to come, we may see other IP rights attaching organically (or at least possibly, with some hoops to jump through) to existing AI-generated trademarks. For now, it remains a serious risk for a business to create an AI company name or slogan (as discussed in Chapter 6 regarding creative writing) and/or then to create a logo for that business using AI, even if it may be technically possible to achieve a trademark registration or benefit from some other IP rights.

4. **Patents – Common, globally**.

Patents are designed to protect inventions rather than artistic creations. Similarly to design rights, which require a functional element, patents can only be secured for inventions that have a real-world use. These types of patents are also commonly referred to as "utility patents." While you can get "design patents" in some jurisdictions, which are broader than applying only to a physical creation, they are broadly intended to cover business interests who have spent time and resources creating something new that will have functional use. The US Patent and Trademark Office defines the distinction as

In general terms, a "utility patent" protects the way an article is used and works (35 U.S.C. 101), while a "design patent" protects the way an article looks (35 U.S.C. 171).[9]

The duration of time that a patent provides protection over the right to benefit and control a unique invention varies between jurisdictions but typically sits around 20 years. This is a relatively long time to be able to exclusively exploit an IP right. When considering that duration, and recognizing that patents are often registered for inventions that may have practical use (therefore might be valuable for someone to buy), patents have the potential to be extremely lucrative assets when in the right hands.

---

[9] USPTO (n.d.) 1502-Definition of a design. www.uspto.gov/web/offices/pac/mpep/s1502.html#:~:text=In%20general%20terms%2C%20a%20%E2%80%9Cutility,171). (Accessed: June 15, 2024)

Patents and AI

It is still unclear whether AI-generated designs or AI-supported inventions (e.g., an inventor using AI to speed up their design or iteration process) is now, or will be in future, capable of benefitting from patent protection.

Of all the regulators or bureaus that govern or determine the validity of IP rights, those with oversight of patents (as opposed to copyrights or other rights) seem to be the most progressive and pragmatic.[10] For example, in early 2024, the US Patent and Trademark Office issued guidance stating that

> ...the inventorship analysis for AI-assisted inventions turns on whether a natural person has significantly contributed to the claimed invention and that the use of an AI system does not, in and of itself, preclude a natural person from qualifying as an inventor...[11]

The UK's equivalent registration body, the Intellectual Property Office, has also held public consultations on this issue. This is indicative of the UK's government increasing recognition that patents are crucial to businesses protecting their innovation and remaining in the UK as a country with a business-friendly rule of law. Mark Westwood (writing in Industry Europe) succinctly summarizes that

---

[10] www.uspto.gov. (n.d.). *USPTO issues inventorship guidance and examples for AI-assisted inventions.* [online] Available at: www.uspto.gov/subscription-center/2024/uspto-issues-inventorship-guidance-and-examples-ai-assisted-inventions.

[11] www.whitecase.com. (2024). *USPTO provides guidance on the patentability of AI-assisted inventions | White & Case LLP.* [online] Available at: www.whitecase.com/insight-alert/uspto-provides-guidance-patentability-ai-assisted-inventions. [Accessed 16 Jun. 2024]

...the government recognises that patent limitations on AI-generated inventions could hinder UK businesses and individuals, and is reviewing their treatment of AI in copyright and patent legislation to seek a balanced solution.[12]

In May 2024, the UK Government released a guidance note which explicitly stated

when the task or process performed by an AI invention makes a technical contribution to the known art, the invention is not excluded and is patent-eligible[13]

What this means in practice is that if real (flesh and blood) human beings have made a substantial, sufficiently clear, and direct contribution to the inventiveness of something, even if AI has significantly supported that process, its patentability will likely be materially similar to a non-AI supported invention process. Encouragingly the odds of successful patent registration in an AI-supported invention process appear to be increasing over time and in an increasing number of countries.

---

[12] Westwood, M. (2022) *Could AI-generated inventions soon be patented in the UK? How with this affect business?* https://industryeurope.com/sectors/technology-innovation/could-ai-generated-inventions-soon-be-patented-in-the-uk-how/.

[13] Guidelines for examining patent applications relating to artificial intelligence (AI) (2024). www.gov.uk/government/publications/examining-patent-applications-relating-to-artificial-intelligence-ai-inventions/guidelines-for-examining-patent-applications-relating-to-artificial-intelligence-ai#what-is-an-ai-invention.

5. **Moral rights – Typical in France, Germany, and many other countries. Rarely seen, used, or enforced in the UK or the United States.**

Broadly, moral rights exist to protect the "honour and reputation"[14] of a creator. Designers (typically artists) have often relied on these rights to

- Claim the right to be named or credited when they create a design and it is displayed, used, or referenced by others. Typically, this can be easily managed by a designer who puts their name on a design (e.g., an artist signing a painting).

- Prevent anyone, other than the designer, from claiming they created the design. Many designers do not want the work of others to be falsely attributed to them, and potentially damage their reputation.

- Prevent others from defacing their original works. For example, making a parody or version of the design which could be distasteful, illegal, or generally derogatory in some way which could damage the reputation of the original design.

In some limited cases, moral rights have also been used against visual creators to protect personal privacy against a creator. For example, if someone is creating a design or taking a picture of something personal,

---

[14] National Portrait Gallery (n.d.) Moral Rights and Artists Resale Right [Accessed: 16 June 2024] www.npg.org.uk/about/creators/moral-rights#:~:text=Moral%20rights%20help%20protect%20a,years%20after%20a%20creator's%20death.

the artist or photographer may be limited in their rights to use the output. So, a photograph of a wedding or private event without the consent of the subjects is likely to have moral rights for the subjects of the photographs.

The use and application of moral rights for this is quite limited. In most jurisdictions, much stronger alternative legal frameworks exist for this kind of privacy protection. Notably, in the United States, personal image rights are more common. In the European Union, individuals can often rely on the European Convention of Human Rights (Article 8 – right to a private and family life).

AI and "Prompt-Only" Creativity

With the types of potential categories of intellectual property in mind, it becomes naturally apparent that designs created via the new era of prompt-based creative processes are at risk of being absent from any protections under current legal frameworks. Broadly, the same considerations apply irrespective of the format of the prompt, for example, verbal commands or text typed into an AI system. In theory, users of complex video and photographic output AI generators such as MidJourney, Sora, DALL-E (and countless more that may soon make these names obsolete) are not only incapable and failing to create something genuinely original, but potentially infringing on IP rights held by those from whom the source material of the underlying models is drawn.

In reality, in the long term, it is unlikely that this will continue to be a material risk. While it cannot be said that there are any certainties for how governments and legal frameworks will adapt to AI in the coming years, it is possible to look at recent history for likely outcomes. We need only look at the history of digitized music distribution to understand the birthing pains of new distribution systems for creative content. Whereas 1999–2012 saw enormous disruption in the music industry, it was *in reality* the birthing pains of a more sustainable long-term distribution system for music. What many people at the time fought over – physical media vs. digital copies, the real battle was over a much larger point – ownership vs. licensing.[15]

In the eyes of most consumers, the physical vs. digital battle over the format of music was futile. The real long-term issue was whether people cared about "owning" music, or simply having access to play what they wanted when they wanted. Streaming has won the battle, and it was decisive. There are certainly a minority who truly care about ownership, and business models remained or even grew to cater for that. In the United States alone, 43 million vinyl records were sold in 2022.[16] This is a small percentage of all the money that consumers paid to listen to music in the same year – $13.3 billion were spent on streaming in the same year[17] – but demonstrates that, over time, the most convenient financial model which has the lowest barriers to use by consumers typically wins when it comes to technology. Governments, regulators, and most businesses understand this.[18]

In 2023, 84% of all music revenue was generated via streaming, while overall music revenue in the United States grew by 10%.[19] This is strong evidence that the early battles over an aspect of how IP rights are respected and enforced (e.g., illegal downloads vs. legal CD and download sales)

---

[15] Rosen, S. (2022) "What makes Spotify tick? An overview of how Spotify licenses music." *Richmond Journal of Law and Technology* [Preprint].

[16] Davis, W. (2024). *Vinyl records outsell CDs for the second year running.* [online] The Verge. Available at: www.theverge.com/2024/3/26/24112369/riaa-2023-music-revenue-streaming-vinyl-cds-physical-media.

[17] Statista. (2018). *U.S. music streaming revenue 2018 | Statista.* [online] Available at: www.statista.com/statistics/437717/music-streaming-revenue-usa/.

[18] www.techuk.org. (n.d.). *Easy does it – the psychology of why we adopt (or reject) technologies.* [online] Available at: www.techuk.org/resource/easy-does-it-the-psychology-of-why-we-adopt-or-reject-technologies.html.

[19] Aswad, J. (2023). *U.S. Recorded-Music Revenue Grows Almost 10% But Vinyl Sales, Streaming Subscriptions Level Off.* [online] Variety. Available at: https://variety.com/2023/music/news/riaa-mid-year-2023-vinyl-streaming-1235725989/.

only matter until the major parties figure out the commercial model which broadly works for businesses and consumers. One could make a strong argument that artists are underpaid in the modern streaming model. However, this is balanced against the reality that streaming is the primary system of music distribution, which does mean that artists can reach a larger audience than ever before (maximizing the chances of artists selling concert tickets), with greater access to nearly unlimited music for audiences. All this, while keeping the large industry-players (in this case, record labels and owners of musical catalogues) sufficiently profitable and satisfied.

A similarly happy medium exploded at the same time with respect to video content, copying much of the model which was pioneered in music – where studios and streaming platforms have too many intents and purposes to become one. If anything, this means that studios such as Netflix, Amazon, BBC, Apple, and many of the legacy studies (some of which merged with the tech-studios) are even more powerful than they were pre-streaming, while accessibility to legal video content for viewers is cheaper than ever (at least by volume). Notably, unlike many of the music streaming services, the video streaming providers combine businesses models and license third party content as well as create their own, and license all of the output to consumers. As with music, one could argue that there are negative consequences for writers, actors, directors, and other creatives – the increased consolidation and power of the streaming providers does mean that creatives are in a weaker negotiating position when selling their services and creations. Nevertheless, a return to the pre-streaming era simply will not happen.

Bringing this back to the legal practicalities, we can see that once the commercial and technological snowball starts gaining irreversible momentum, the law and business combine to catch up. While this is not a certainty, and there may be differences between different mediums (e.g., written text may end up being treated differently to a creative or utilitarian design), it is highly probable that the risks of using AI to create designs will diminish over time and such use will become a norm (if not already).

The Risk to a Designer Using AI

Almost inevitably, in the first few years of the AI-powered creative boom – there will be lawsuits, lots of lawsuits. We can already see this with the New York Times v OpenAI proceedings which center around the monetization of outputs drawn from models containing allegedly unauthorized data (which may be protected intellectual property).[20] However, once some time has passed, it is inevitable that businesses will move faster than legislators to settle on commercial models that work for all involved (or at least the big players with the most influence).

Practically, in these early days of mass AI-powered creativity adoption, this means that users might need to worry somewhat that they risk being caught in the crossfire of these IP battles. In the same way that many of the participants in illegal music downloads were also implicated in criminal cases,[21] rather than just the companies facilitating the downloads, such as Napster or The Pirate Bay,[22] there is always a real concern that for many – that potential risk outweighs the benefits. This is a real, mostly untested, risk. There is no sugarcoating the possibility that if a designer were to create something using an AI tool which has replicated the works of another person or company without their permission, they may face direct legal actions from the aggrieved party.

---

[20] OpenAI claims New York Times "hacked" ChatGPT to build copyright lawsuit. (2024). *The Guardian.* [online] 27 Feb. Available at: www.theguardian. com/technology/2024/feb/27/new-york-times-hacked-chatgpt-openai-lawsuit#:~:text=Representatives%20for%20the%20New%20York [Accessed 14 Apr. 2024].

[21] Holpuch, A. (2012). *Minnesota woman to pay $220,000 fine for 24 illegally downloaded songs.* [online] the Guardian. Available at: www.theguardian.com/ technology/2012/sep/11/minnesota-woman-songs-illegally-downloaded.

[22] Goldberg, D. and Larsson, L. (2017) "Pirate Bay co-founder Peter Sunde: 'In prison, you become brain-dead," *The Guardian,* 21 February. www.theguardian.com/technology/2014/nov/05/ sp-pirate-bay-cofounder-peter-sunde-in-prison.

In the long term, however, the risks will almost certainly diminish for the ordinary designer using an AI system. It is now extremely rare for individuals to face prosecution for illegally downloading music or films, despite the theoretical legal consequences remaining significant.[23] Considering we are several years into the age of sophisticated AI-generated images, video, and music (i.e., an era where the quality is sufficient that creators actually want to use the technology), there have been surprisingly few legal claims against creators who use the AI systems. The reality is that it would only make sense, whether for music piracy or AI-supported infringement of other people's creations, to take a legal action against the creator who uses an AI system if

1. The goal is simply to limit the damage which may already be done by the accused, that is, seeking an injunction (the support of the court to prevent further use or misuse) or

2. The accused has significant financial resources (typically a business, usually with insurance) to pay compensation

It is often incredibly hard, practically, for the owner of original IP that is being used as part of the source data in a large model to actually prove their content is being used/misused. On the other hand, it is equally hard for a creator using AI to know exactly what AI "creativity" went into each pixel of an image, frame of a video, or note of music vs. how much of it was simply copied from another party. This makes enforcement (particularly compared to the illegal free downloading of music and video) significantly more challenging. So, creators are likely (though not certain) to be able to create using AI systems with relative impunity until better transparency in underlying models becomes the norm.

---

[23] Stuart Miller Solicitors. (2024). *Want to know the sentence for a digital piracy copyright offence in 2024?* [online] Available at: www.stuartmillersolicitors.co.uk/sentences/sentence-for-digital-piracy-copyright-offence/.

The Dark Side of Using AI for Creativity

The legal risks of using AI are not just rooted in IP. They can extend to other equally or more serious, even criminal, issues. Most notably, so called "deepfakes" have become a real legal issue for creators of photo and video imagery to navigate when considering the content and subject of their creative output, as well as the intended audience. Not all deepfakes are created equally, whether in quality or intention. However, most deepfakes at least have the potential to be damaging in unpredictable ways – which can have potential legal consequences.

Many would consider the most "safe" and commercial applications of deepfakes to be the use of a person's image or voice layered onto another person's body or onto a digital creation for the purposes of film, TV, audio, and art. However, this is a hugely divisive issue. Not only does it potentially create a significant detrimental risk to actors (and voice actors) by allowing commercial third parties to use their likeness or voice without ongoing consent, but encourages studios and producers to push artists of all kinds to sign deals assigning their rights or granting permission for such deepfakes to be legally possible. This would be most likely to happen when a performer is early in their career and have the least bargaining power, and their likeness is the least expensive to acquire. Not everyone agrees that this is inherently a bad thing, with some arguing that it may allow actors to be paid for work they don't even have to physically attend or beyond their ordinarily expected career duration.[24] For studios, financiers, and producers behind creative projects, it could also be said that such capability gives them greater certainty that a project will reach completion e.g., in the event that a performer is unable to continue due to health,

---

[24] Nolan, B. (2023). *Some actors and celebrities say the popularity of AI deepfakes could be good for their careers.* [online] Business Insider. Available at: www.businessinsider.com/ai-deepfakes-actors-strike-writers-careers-2023-3 [Accessed 16 Jun. 2024].

personal issues, or even working conflicts. Legally, image rights (as already discussed in this book) are not a globally recognized legal right. However, in most jurisdictions it would be legally/contractually possible for a person to make a deal to allow the use of their face or voice by a third party if that third party is paying them and for a defined duration, within specified use cases. For example, the use for 1 year of a person's voice to carry out digital edits to a voiceover. Whether the deal is fairly paid will be fact-specific, but it is clear that practically, the power of deepfakes gives the "big players" of the media industry potentially significant leverage against performers.[25]

If a digital designer, for example, a person working in visual effects or sound engineering, is being asked by a studio or production company to use someone's image or voice as part of a digital insertion into a visual or audio creation, it is advisable in most cases for the designer to ask whoever has commissioned them to provide some kind of written assurance or evidence that they are doing so with the consent of the person whose likeness is being replicated.

Whether a photo or video is *intentionally* a deepfake, and whether it is intended to deceive or simply be a new creative piece, may be debatable in some circumstances. However, irrespective of the intention of the creator/ designer, they may face legal challenges (either civil or criminal, or even both – depending on jurisdiction) if the deepfake

---

[25] Beckett, L. and Paul, K. (2023) "'Bargaining for our very existence': why the battle over AI is being fought in Hollywood," *The Guardian*, 24 July. www.theguardian.com/technology/2023/jul/22/sag-aftra-wga-strike-artificial-intelligence.

1. Risks compromising or damaging the reputation, privacy, or personal/work life of the subject

2. Potentially implies endorsement of something, or fraudulently states something on behalf of the subject

3. Breaches a direct and specific legal safeguard, for example – the Online Safety Act ratified in 2023 in the UK[26]

Some deepfakes are not full video, photo, or audio – but may be manipulations of existing/live content. For example, there are AI-powered applications which make it appear that a user's eyes are always facing the camera and paying attention even when they are looking off-screen or have their eyes shut.[27] While applications like these may seem harmless, they can be misused depending on one's perspective. Although most countries do not have a mandatory legal requirement to disclose when using such technology, an employer carrying out a video-conference interview of an interviewing candidate may feel misled in the hiring process.

Practically, more companies are asking their employees (or prospective employees) not to use AI in the hiring process as well as in many aspects of their day-to-day roles where that usage may make someone else in the business or another business/customer feel misled or uncomfortable. On the other hand, many people would consider that if a business is using AI

---

[26] Ministry of Justice (2022). *New laws to better protect victims from abuse of intimate images*. [online] GOV.UK. Available at: www.gov.uk/government/news/ new-laws-to-better-protect-victims-from-abuse-of-intimate-images.

[27] RisingMax (2024). *AI Eye Contact App Development | Launch AI Eye Contact Tool*. [online] risingmax.com. Available at: https://risingmax.com/product/ ai-eye-contact-app-development#:~:text=Eye%20Contact%20AI%20can%20 help [Accessed 16 Jun. 2024].

systems as part of the hiring process to assess engagement through metrics such as eye contact, this should also be disclosed. In some countries, that type of analytical technology may be considered legal high-risk or even illegal.[28]

Weaponization of Deepfakes

There is a rapid move in many jurisdictions to limit the non-consensual creation of deepfakes using the image of real people. Aside from the commercial applications already mentioned, there is an even darker side to deepfakes. The UK was one of the first to move on this legislative guardrail, but many other countries are following suit.[29] Without needing to linger on the obvious moral issues (i.e., creating an image of someone doing something, which they didn't, is generally distasteful and wrong on many levels), there are likely to be serious legal penalties for the creators. The consequences could (and rightfully should) be more severe where a deepfake has created a false perception of someone doing something highly compromising or which may affect their reputation – the most common example being deepfake pornography (where a person's face is manipulated onto another's body). Notably, the United States has been slow to federally regulate deepfakes per se. However, many existing federal and state laws already regulate the impact of this phenomenon – fraud,

---

[28] Welker, Y. (2024) "Can the EU AI Act embrace people's needs while redefining algorithms?" Euronews, 29 January. www.euronews.com/my-europe/2024/01/29/can-the-eu-ai-act-embrace-peoples-needs-while-redefining-algorithms.

[29] Nowell, T. (2024). *New laws criminalise the sharing of intimate deepfakes without consent.* [online] Refuge. Available at: https://refuge.org.uk/news/new-laws-criminalise-the-sharing-of-intimate-deepfakes-without-consent/.

misinformation, and/or invasions of privacy. Encouragingly, the EU's AI Act 2024 has specific provisions regarding the use and misuse of deepfakes. Essentially, making deepfakes illegal in many circumstances and always when they are not clearly disclosed as being AI-generated.[30]

In the long term, deepfakes create a real erosion of public trust in news and public figures, as well as creating enormous risks of identity fraud and scams.[31] Without even touching on the significant risks to political transparency and voter confidence (there are many good books on this subject – Deepfakes and the Infocalypse by Nina Schick is highly recommended), deepfakes have some useful commercial applications and significant societal risks. Countless examples already exist of fraudsters using deepfakes as a way to con real people into desperate situations.[32] In addition, those individuals whose faces are misused as part of these scams can have their reputations ruined if they are not quick to act and control the narrative. In the UK, a prominent "money saving expert," Martin Lewis, was the subject of a deepfake to promote a scam,[33] causing

---

[30] Atsüren, A. (2024) "EU AI Act 2024 | Regulations and Handling of Deepfakes - BioID," *BioID*, 7 June. www.bioid.com/2024/06/03/eu-ai-act-deepfake-regulations/#:~:text=Aspects%20of%20the%20EU%20AI%20Act%20Regarding%20Deepfakes&text=Developers%20and%20users%20of%20deepfake,the%20content%20they%20are%20viewing.

[31] Vaccari, C. and Chadwick, A. (2020) "'Deepfakes' are here. These deceptive videos erode trust in all news media.," *Washington Post*, 27 May. www.washingtonpost.com/politics/2020/05/28/deepfakes-are-here-these-deceptive-videos-erode-trust-all-news-media/.

Owen, A. (2024). *Deepfake laws: is AI outpacing legislation?* [online] Onfido. Available at: https://onfido.com/blog/deepfake-law/.

[32] Tracey, M.D. (2024). *Scammers Use Agent Deepfakes to Fool Buyers, Sellers.* [online] Available at: www.nar.realtor/magazine/real-estate-news/technology/scammers-use-agent-deepfakes-to-fool-buyers-sellers.

[33] Shaw, G. and Lekarski, P. (2023). Beware frightening new "deepfake" Martin Lewis video scam promoting a fake "Elon Musk investment" – it's not real [online] Available at: www.moneysavingexpert.com/news/2023/07/beware-terrifying-new--deepfake--martin-lewis-video-scam-promoti/.

potential mistrust in his (otherwise generally positive) reputation. The harm to the victim of a deepfake scam is likely to correspond to the level of trust they would reasonably place in the subject of the deepfake, making this particular scam particularly damaging. Legally, we can expect greater restrictions across the globe akin to the minimum transparency requirements already seen in the EU AI Act 2024, with some countries potentially legislating for outright bans. The latter may be very difficult to enforce in practice.

At present, many social media providers do not consider it their explicit duty to regulate or control the use of deepfakes on their sites unless they breach other guidelines (e.g., are obviously criminal or, in some cases, pornographic). So, a potential user of photo or video generation using AI should remember that intentionally creating a deepfake with the intention to harm the subject, or mislead the viewer, is likely to be on the wrong side of the law (at least in the long term) while also the perpetrator of a significant ethical harm.

Accidental Creation of Deepfakes

What is concerning for many creators is that they may not realize they are creating a deepfake drawing on the image of real people, when they are creating an AI-generated or AI-assisted design (most likely in a generated "photo" or video). While this risk in most circumstances would be low, it is not zero.[34] It would be legally challenging, embarrassing, and potentially expensive to address should a designer generate a video, photo, or audio creation which misuses someone's likeness without their permission, irrespective of intent.

---

[34] Tech firms sign "reasonable precautions" to stop AI-generated election chaos. (2024). *The Guardian*. [online] 16 Feb. Available at: www.theguardian.com/technology/2024/feb/16/tech-companies-precautions-ai-election.

So, creators should take reasonable steps to ensure that when using AI tools to generate images or videos, that there is a clear understanding of the source material (if at all possible). Practically, this means using the providers of AI systems which give the greater assurances and have easier to understand terms and conditions – transparency is typically a positive. Or, at the very least, make some basic checks as to whether something which could be considered photo-realistic looks like anyone famous (easier said than done). It may sound silly, but even some evidence that some basic google-searching of an image to see if anything matches could be helpful should there be any later defense required of the creation. Ironically, there are even AI tools which can help search for whether someone's image looks like a celebrity or an image online, and these should not be overlooked for their potential practical value. Many AI tools to create photo or video content already may have built-in guardrails against deepfakes,[35] but whether these are robust in the long term remains to be seen.[36]

Key Considerations

1. As with creative writing, AI can be used safely, legally, and in a relatively low-risk and low-cost manner to accelerate the processes of design of many kinds, from simple static images to complex videos.

---

[35] Hsu, J. (2024). *Realism of OpenAI's Sora video generator raises security concerns.* [online] New Scientist. Available at: www.newscientist.com/article/2417639-realism-of-openais-sora-video-generator-raises-security-concerns/.

[36] Jackson, B. (2024). *OpenAI's Sora Indicates How Deepfakes Are About to Worsen.* [online] Spiceworks. Available at: www.spiceworks.com/tech/artificial-intelligence/guest-article/deepfakes-are-about-to-become-a-lot-worse-openais-sora-demonstrates/.

2. When using an AI system to generate an image or video, it is very hard to know exactly where every image has been drawn from/recreated from. So, it is reasonable to exercise particular caution when generating images or videos with faces, and audio which sounds like a real voice.

3. While the misappropriation and misuse of words can be seriously damaging, the potential damage of a misleading (or even fraudulent) image, video, or piece of audio is often much more impactful and harder to undo. So, it is important to consider two major ethical questions:

    a. Do I believe that I am doing something wrong by potentially copying the imagery, video, themes, or previously original visual and design concepts from someone else?

    b. Does my design output, whether a still image, video, or otherwise, potentially risk misleading the viewer, including as to whether AI was used in its creation?

4. As a creative user of AI systems, quite simply you cannot be fully certain that your output design fully excludes content that infringes the legitimate IP rights of other designers or creatives. However, your risk *may* be lower than with the often more obvious infringement that arises around the misappropriation of words. A sentence that is an exact copy of another is a clearer risk than an image which looks similar to another.

5.   As with the use of AI for writing, there is always likely to be *some* potential legal claims you could face if, in practice, your design output is somewhat a copy of another's work without their permission. Again, your copying is still likely to be considered copying in the eyes of the law irrespective of whether the AI provider was negligent or themselves in breach of the law in allowing you access to the source material.

6.   The law is even more uncertain with respect to visual and audio design than the written word. Current and future legal cases as well as recent and potential legislation may have a significant impact on the level of risk posed to design creators who are using AI systems.

7.   While the legal risks are greater when using AI tools compared to the creation of an entirely original piece of visual or audio design without the help of AI, it is not a realistic long-term approach. AI is here to stay. The law in many jurisdictions is just starting to catch up and will in future give greater certainty to well-intentioned creators.

CHAPTER 7

AI for the "Professions"

As a young Jewish child growing up in North London, I was under the impression that the right answer to the frequently asked "What do you want to be when you grow up?" only had three possible correct answers:

1. Doctor

2. Lawyer

3. Accountant

Mostly, because those were the only three careers I knew of beyond that of my dad who was a textile merchant. Although the textile trade is probably less vulnerable, though not impervious, to the impacts on businesses stemming from AI – the same cannot be said of medicine, law, or accountancy. With hindsight and a long-term perspective, it is clear that physically involved careers such as the "trades," for example, plumbers, electricians, builders, remain resilient and buoyant globally, irrespective of AI adoption in business.

This is the longest chapter of this book for a good reason. The professions are the sector where the application of AI is arguably the most complex and, in many cases, with the highest direct risks. Some of these risks apply to the professional service providers, others to individuals outside of the professions (i.e., ordinary members of the public, untrained as doctors, lawyers, or accountants). Non-professional individuals may be

© Harry Borovick 2024
H. Borovick, *AI and the Law*, https://doi.org/10.1007/979-8-8688-0400-7_7

end clients of professional service providers using AI, but may also be (by the very nature of how AI tools can be used) directly accessing AI-powered tools applied to professional fields on a self-service basis. This chapter will focus on understanding these professions, then how they will be transformed by AI, and then how this will practically impact, empower, or be useful to ordinary people who use or need these professional services.

The "Professions" vs. Other Roles

While the day-to-day of these jobs, and other similar highly skilled professions such as architecture or engineering, can vary wildly – they share a common thread. Of course, being "a professional" can apply to any career. However, each of the professions considered in this chapter, with varying levels of risk and regulatory obligations, owe some kind of legal duty to the recipient of their work. Typically, accreditation is required for each of these roles and these jobs are regulated, with specific legal frameworks that govern what is expected from individuals and organizations within those professions. The UK government actually maintains a list of every job they consider to be a regulated profession.[1] As considered in Chapter 5, many financial services are also regulated and providers owe a duty of care to their clients, but the extent and enforcement of that regulatory obligation varies significantly based on seniority of the individual concerned as well as the type of financial service. These regulatory obligations can even apply to conduct outside of the workplace – that is, a lawyer could lose their right to practice law if they are found to have behaved inappropriately (typically criminally) even

---

[1] Gov.UK (2021). *UK regulated professions and their regulators.* [online] GOV.UK. Available at: www.gov.uk/government/publications/ professions-regulated-by-law-in-the-uk-and-their-regulators/ uk-regulated-professions-and-their-regulators.

if it has little direct relation to their work.[2] The reason for this is that the professions depend on one thing more than anything else – trust, from their clients and from the public.

The focus of this chapter (mostly for simplicity) is on the three specific professions of law, accounting, and medicine, but should be broadly applicable to any professional operating within an established and regulated industry where there is a personal level of direct regulatory obligation (i.e., responsibility isn't necessarily shielded or borne by the employer but can also have personal repercussions for the individual practitioner).

Good Practice

Each of these regulated professions share, with the exception of emergency hands-on medical care carried out by medical professionals, that they are typically advisory in nature – at least initially. For example, even in the case of surgeons, their role is often not just to operate but also to exercise judgment and advise the patient whether the surgery is appropriate. There is also at least some objective understanding of what is good practice for a practitioner within each of these roles. Typically, this would be reasonable quality of well-considered advice which allows the end client to take informed action or inaction. In the United States, the expected quality of competence is that of "a well-qualified professional acting under the circumstances presented."[3]

---

[2] Klevens, S. L and Clair, A. (2021) Dentons. Available at: *Can lawyers get in trouble with the bar for what they do in their private lives?* www.dentons.com/en/insights/newsletters/2021/march/10/practice-tips-for-lawyers/can-lawyers-get-in-trouble-with-the-bar-for-what-they-do-in-their-private-lives.

[3] MacEntee, B.F. and Schnake, K.G. (2023) In review: professional negligence law in USA. https://lexology.com/library/detail.aspx?g=6f9252da-df48-4c9a-949a-4aef94e735a2.

It's important to note, this is a *reasonable* quality of professional care and it is unlikely that any regulator would expect to hold a professional to a *high* standard of care. This may seem dangerous to an individual on the receiving end, but in reality, it makes operational and practical sense. The law in most countries recognizes that no profession operates under perfect circumstances, even if they are well paid. Rather, most professionals are expected to meet the standard that their peers would reasonably achieve.

Bad Practice

There is also some real idea, varying from country to country, of what poor or even negligent performance by a professional might be. This would be typically determined and enforced by some kind of regulatory body at a regional or national level. While poor performance can be reputationally damaging, genuine negligence in any of these professions can have serious legal consequences, ranging from suspensions and exclusions from the right to practice that role to prison time. The human reality of negligence on the client (who at this point may be considered a victim) can also be extreme. For obvious reasons, medical negligence is often the most detrimental and serious professional negligence one person can (whether purposefully or accidentally) inflict on another. Even negligence by an accountant or lawyer can have a hugely detrimental effect on the life of the client, for example, when a false tax filing is made that results in potentially criminal ramifications.

In British law, the standard of professional care in medicine (for example) is called the "Bolam" test. You don't need to remember that, but it is important to bear in mind that the expected standard of care before something can be considered clinical negligence varies between countries. The UK standard under the Bolam test is whether the act or practice by the doctor would be supported by a responsible body of similar professionals.[4]

---

[4] Samanta, A. and Samanta, J. (2003). Legal standard of care: a shift from the traditional Bolam test. *Clinical Medicine*, [online] 3(5), pp. 443–446. doi: https://doi.org/10.7861/clinmedicine.3-5-443.

This is clearly a lower bar than the US standard already mentioned. This UK approach is imperfect and often criticized. Many countries, not just the United States, take harsher or a more relaxed approach, so a client's rights and expected standard of care may vary significantly based on their country of residence or the country in which the service is performed. In the United States, the test whether a professional has acted reasonably is typically more client friendly[5] mostly, as US professionals carry far greater insurance to pay out to disgruntled clients or even those who have experienced some kind of serious harm.

It should be noted that *regulated* individuals and companies providing financial services can also be considered within this same scope (see Chapter 5). However, it is possible to provide some financial services and act in some investment capacity in many jurisdictions without a license from a regulatory body. On the other hand, the practice of law, medicine, or accountancy in most jurisdictions without a license may be a civil or even criminal matter with serious legal consequences. So, before considering whether AI can perform the professions, it's worth recognizing that

1. For now, there is no AI system which is regulated and authorized to give direct medical, legal, or accounting advice.

2. Only human beings are currently recognized by regulators as being capable being regulated and appropriately giving advice/practicing these professions.

---

[5] Liddell, K. *et al.* (2022) *DIFFERENTIATING NEGLIGENT STANDARDS OF CARE IN DIAGNOSIS, Medical Law Review*, pp. 33–59. https://doi.org/10.1093/medlaw/fwab046.

3. Without a regulated individual or organization
    providing the regulated activity, it's unlikely that
    a user of any similar services provided by an AI
    will benefit from any of the same safety nets which
    would otherwise apply.

Can AI Perform the Professions?

*The world is changing, and in that changing world, expecta-
tions are growing. In many situations AI can help meet those
expectations. Medicine is no exception. However as a generation of physicians who strongly leant to the professionally artistry view of medicine, I do not think we have seen the end of
the doctor. There is a role for AI, we (the Profession) need to be
involved and engaged in its development. But remembering
that human interaction is part of being human.*

*Perhaps the younger generation are moving faster towards a
technical rational view of the world.*

Is this good or bad... time will tell.

—Dr. Tim Battcock, MBChB FRCP (Consultant Physician)

AI will not fully replace the professions in our lifetimes. This might
seem like a bold statement considering the rapid advancements and
applications of AI, but it's rooted in one reality – people want another
person to trust, or to blame. It's arguable that the "person" might be
the providers of AI systems (i.e., the people owning and running AI
companies), and to some extent that will happen. However, digital tools
to accelerate delivery, quality, and accuracy of professional services are
nothing new.

Doctors have long used robotic arms and indirect instruments to assist in surgical procedures,[6] or even simple video conferencing tools for remote consultations and second opinions, since the early 2000s. AI is likely to be most immediately useful in medicine as a diagnostic assistant[7] as well as an initial touch point for patients with common queries. In the long term, drug discovery within the pharmaceutical industry is also quite promising with companies such as Benevolent AI achieving very large fundraises to accelerate medical research and bringing new products to market.[8]

Less glamorously, but with extreme pervasiveness, accountants have been using Microsoft Excel and similar spreadsheet platforms for decades in addition to more specialized accounting and bookkeeping software. Lawyers have historically been one of the slowest professions to digitize and adopt new technologies. However, the AI era is different. Lawyers are at the forefront of adoption of AI-based technologies in the current AI boom.[9] With a multitude of potential use cases, including legal research on cases and legislation,[10] gaining insights into large volumes of documents,

---

[6] *Robotic surgery - Mayo Clinic* (2024). www.mayoclinic.org/tests-procedures/robotic-surgery/about/pac-20394974#:~:text=The%20most%20widely%20used%20clinical,view%20of%20the%20surgical%20site.

[7] Marr, B. (2024) "How generative AI will change the jobs of doctors and healthcare professionals," *Forbes*, 14 March. www.forbes.com/sites/bernardmarr/2024/03/13/how-generative-ai-will-change-the-jobs-of-doctors-and-healthcare-professionals/#:~:text=AI%20As%20A%20Diagnostic%20Assistant&text=This%20will%20lead%20to%20more,%2C%20nurse%2C%20consultant%20or%20specialist.

[8] *BenevolentAI announces $90 million investment from Temasek* (no date). www.benevolent.com/news-and-media/press-releases-and-in-media/benevolentai-announces-90-million-investment-temasek/#:~:text=%E2%80%8D,a%20Singapore%2Dheadquartered%20investment%20company.

[9] LexisNexis (2021) Lawyers cross into the new era of generative AI. www.newlawjournal.co.uk/content/lawyers-cross-into-the-new-era-of-generative-ai.

[10] Stokel-Walker, C. (2023) "Generative AI is coming for the lawyers," *WIRED*, 21 February. www.wired.com/story/chatgpt-generative-ai-is-coming-for-the-lawyers/.

automated contract negotiation,[11] and even to review evidence in court cases,[12] there seems to be little room to hide from AI in law. Arbitrators and even judges increasingly recognize that AI is having, and will continue to have, a wide and deep effect on the administration and determination of justice within the frameworks of both criminal and civil law.

A 2023 guidance note, published by the UK's judiciary to help guide judges understand the risks and advantages of AI, made it clear and unambiguous as to where the burden of responsibility lies:

> *The accuracy of any information you have been provided by an AI tool must be checked before it is used or relied upon.*[13]

This may seem like a simple statement, but it concisely contains the fundamental principle that underpins the core message of this book – with a particularly important impact on the professions as regulated industries. A doctor, accountant, or lawyer faces real consequences when making poor or negligent decisions. Verification and critical thinking are core skills within each of these professions – whether of text books, the advice of others, or outputs (such as recommended courses of action) from AI. A patient will generally not expect a regulator to allow a doctor to escape real career ramifications (including direct financial compensation) should they provide negligent medical advice by virtue of it being from AI. This expectation applies even if that AI is verified by other professionals than the one from the end client is taking the advice. So, even the positioning by a professional that they trusted an AI output as it is produced, provided,

---

[11] McManus, S. (2023) Can AI cut humans out of contract negotiations? www.bbc.co.uk/news/business-67238386.

[12] Cross, M. (2023) *News focus: Artificial intelligence debuts at the Old Bailey.* www.lawgazette.co.uk/news-focus/news-focus-artificial-intelligence-debuts-at-the-old-bailey/5114799.article.

[13] Artificial Intelligence (AI) Guidance for Judicial Office Holders 12 December 2023, Courts and Tribunals Judiciary, Available at: www.judiciary.uk/guidance-and-resources/artificial-intelligence-ai-judicial-guidance/.

or reviewed by other professionals is unlikely to be a sufficient defense for their own failure to verify and provide reasonably competent advice and services. Accordingly, a severe lack of clear practical guidance by many national and regional regulators as to how professionals should use AI systems responsibly and safely has drawn intense criticism from practitioners.[14]

The same applies to other professions where a certification or license is required by a governing body with regulatory oversight. For lawyers, accountants, and doctors, there will be some variation between jurisdictions. However, as an example, a lawyer's advice to their client is just that – it will be regarded as the advice they have provided with the expectation that it can/should be relied upon by their client, irrespective of how it was devised or produced. This means that no amount of "the AI told me so…" is likely to wash with a regulator (or client, for that matter) should there be a serious error in the advice.

In reality, the impact of legal or pseudo-legal advice and commentary is already visible outside of the direct lawyer-to-client fiduciary relationship, notably in business-to-consumer interactions. A Canadian court determined in early 2024 that a chatbot which gave inaccurate information to a consumer should be held to the letter of its outputs,[15] even if the information provider does not owe a strict legal duty (i.e., they are not that person's legal advisor). In this case, a customer was given inaccurate information regarding discount fares for travel with Air

---

[14] Smith, H., Downer, J. and Ives, J. (2023) "Clinicians and AI use: where is the professional guidance?" *Journal of Medical Ethics*, p. jme-108831. https://doi.org/10.1136/jme-2022-108831.

[15] Lifshitz, L. R and Hung, R. (2024) "BC Tribunal Confirms Companies Remain Liable for Information Provided by AI Chatbot," *American Bar Association* [Preprint]. www.americanbar.org/groups/business_law/resources/business-law-today/2024-february/bc-tribunal-confirms-companies-remain-liable-information-provided-ai-chatbot/#:~:text=In%20Moffatt%20v.,policy%20for%20discounted%20bereavement%20fares.

Canada, by the airline's own chatbot. What this meant legally was that the chatbot (and therefore Air Canada) had made a negligent misstatement or misrepresentation to the consumer.[16]

Of course, a chatbot is not currently considered in any major jurisdiction to have any legal personality (it's not a person or a company). Accordingly, it cannot be held to any kind of legal account on its own. So, the company who provide and publish the chatbot should be deemed legally responsible. It's worth noting that Air Canada actually did try to specifically pursue the defensive line of argument in court that the chatbot was its own entity, but this was rejected outright, providing a useful and practical precedent. Simply – we know that it doesn't matter if there is static text, or complex interactive services, it is the provider/publisher who bears legal responsibility of the information or advice they give out to the public (i.e., the person or company from which the end customer believes they are receiving the information).[17]

It's easy to imagine the consequences if this wasn't a misstatement regarding something as relatively mundane as a discount policy. Imagine instead taking this to reasonably probable scenarios for AI-powered chatbots or other interactive services

- Incorrectly prescribing or rejecting the prescription of medication.

- Giving false or outdated tax advice.

- Advising someone they had a legal right or justification (or didn't have a legal right) to something when the opposite was true. This would be particularly impactful in matters of criminal law.

---

[16] Yagoda, M. (2024) *Airline held liable for its chatbot giving passenger bad advice - what this means for travellers.* www.bbc.com/travel/article/20240222-air-canada-chatbot-misinformation-what-travellers-should-know.

[17] Garcia, M. (2024) "What Air Canada lost in 'Remarkable' lying AI chatbot case," *Forbes*, 20 February. www.forbes.com/sites/marisagarcia/2024/02/19/what-air-canada-lost-in-remarkable-lying-ai-chatbot-case/.

The personal consequences to the end customer (particularly in the case of any medical misstatement) could be extreme, and therefore is the liability of the provider of these interactive services. In any case, the reason why liability and regulatory obligations exist with respect to these professions is because the consequences of poor delivery can be life-changing and unpredictable.

Having said all of the preceding, we should not underestimate just how powerful AI tools can be for each of these professions. It would also be a mistake to underestimate how common it will be for practitioners to use AI regularly in the imminent future. Clients already may, and certainly will, expect the professional, on whom they are placing reliance, to use all reasonably available resources to improve the accuracy, quality, speed, and overall usefulness of their advice.[18] Arguably, as AI systems become a typical tool in each profession, failure to appropriately use AI systems to provide quality advice in a time-efficient manner could eventually be considered a failure to meet the reasonably expected standards of care.[19]

Principles As a Practitioner

The chances are that if you are reading this book as a lawyer, accountant, or doctor – you didn't have even the slightest of training during the qualification process as to how AI can accelerate, improve, or streamline professional practice. The professions have been particularly slow to incorporate AI-related content as part of core, or even optional, training.

---

[18] Crestodina, A. (2024) *AI powered services,* Orbit Media, www.orbitmedia.com/blog/ai-powered-services/.

[19] Houlston, H. (2024) *Considering professional negligence in the use of AI - whose fault is it anyway?* www.ashfords.co.uk/insights/articles/considering-professional-negligence-in-the-use-of-ai-whose-fault-is-it-anyway#:~:text=Effective%20engagement%20with%20AI&text=Given%20the%20potential%20of%20AI,employ%20AI%20within%20their%20practice.

By contrast, some educational frameworks which often lead people into regulated professions (e.g., business schools) have been notably faster to move and adopt forward-looking AI education.[20]

One of the reasons for the slow uptake of technological education within these professions is the unusually slow adoption of technology in practice compared to many other industries. Although technology may be useful and found in some of the most sophisticated medical, accounting, and legal organizations, most are relatively antiquated and working at the level of minimum viable requirements. For example, in the United States, the use of fax machines remains common for communication within and between hospitals.[21] There are also tremendous global disparities with respect to available resources and training.

It is also critical to remember that some of the most useful skills for professional practice are social (e.g., empathy from a doctor) rather than technical or technological. As there is only a finite amount of time which can be dedicated to professional education, it is hard to strike a balance between technical vs. soft skill vs. technology training. The nuance here is that AI skills may free up time for the rest of the professional's training and actual practice.

> *The integration of AI into clinical practice is inevitable. In the same way that the (ongoing) digitalisation of healthcare has facilitated medical practice, AI will also become a useful tool. It will facilitate rather than replace the clinician.*

---

[20] Ellis, L. (2024) *Business Schools Are Going All In on AI*. Wall Street Journal, www.wsj.com/tech/ai/generative-ai-mba-business-school-13199631?st=4n7z8trrqg5gwqm.

[21] Segal, B. (2023) "Here's why hospitals still use fax," *Telnyx*, 29 September. https://telnyx.com/resources/why-hospitals-use-fax.

If used effectively this will free up more time for doctors to do what AI cannot. Blowing bubbles to soothe an anxious toddler; ethical discussions around consent or the withdrawal of care; holding the hand of frightened patient at the end of their life.

The human element in both diagnosis and treatment is central to clinical work. If AI can provide doctors with more time at the bedside, that can only be a good thing. (Though a fully digitalised system will be a necessary first step!)

—Dr. George A.G. Williams, NHS Doctor

So, with this context in mind, it's important to understand three key principles as an AI-powered (or even slightly AI-assisted) professional:

1. **If you are regulated – it's your regulatory responsibility.**

2. **If you don't know where the data goes, or who can see it, you're taking a bigger risk.**

3. **Usefulness is more important than intelligence.**

Looking at the detail of these principles, there is quite a lot to unpack with considerable nuance. However, we can also see very practical ways in which AI can be used safely and with appropriate caution while still delivering great potential upsides.

If You Are Regulated – It's Your Regulatory Responsibility

The reality is that regulators are unlikely to care, in most circumstances, what tools were used by a professional in the course of daily practice so long as they can be considered among peers within that profession as reasonably safe to use. The obvious exception to this would be the choice and usage of hardware applied by medical practitioners on patients. What this means is that if it all goes wrong (e.g., negligent advice is given, or poor-quality action is taken), the human doing the work at the point of delivery is where the buck stops. There is no real way to dance around that direct consequence.

Although a regulator may take extenuating or specific circumstances into consideration, it is clear that dependency on AI without additional verification of a sufficient quality just simply isn't good enough. Crucially, this goes to the core of the rationale for using a professional in the first place. An untrained individual outside of a regulated profession should reasonably expect to have a professional service and level of client care on which they can place reliance.

A solid basis of client care, including or excluding the use of AI or any other technology, can be defined as: *a reasonable quality of consideration (thought and effort) + reasonable quality of verification + a reasonable quality of application.*

The trust in, and longevity of, professional services is dependent on the interface between the end customer and the professional. For example, as per some of the perspectives quoted in this chapter, consider the dynamics of a patient's relationship with their doctor or nurse. It is typically important that there is foundational trust by a patient in the advice they receive. While it is not uncommon for patients to seek second opinions, that still reflects some overall (or at least basic) trust in the medical practitioners as a whole. In this particular case, if a doctor is using an AI tool to assist with a diagnosis, it would be sensible to think of the AI as the first (or partial) diagnostic opinion and the human as the second opinion who is able to take the helpful aspects of the initial opinion while adding their own experience and additional patient context to deliver a more complete patient-facing opinion. Even if the AI is used as a verification of a doctor (i.e., where the AI is the second opinion), this does not absolve the doctor or hospital/clinic of their own responsibility.

Usefulness Is More Important Than Intelligence

Publicly available chatbots such as ChatGPT are widely available. However, one of the fastest commercial growth areas in the use of AI by professionals is pre-trained specialist chatbots.[22] These are used either (1) to answer the queries of the professionals themselves; or (2) to answer queries of other people, to which the professional would typically answer.

The reason for swift growth of such tools is simple. Either

1. They fill a repetitive, manual, and often low-skill aspect of the workload typically borne by highly skilled professionals

   or

2. They serve to remind, refresh, or update the professionals regarding specific information that they do not keep to hand, minimizing the need to search or organize

Within a business, it will often be the cases that legal and accounting will be core users of such tools, but their usefulness often extends into the realms of information technology and security, insurance, human resources, and compliance.

A typical practical use case of such tools is using the outputs of the chatbots in the population of RFPs (requests for proposals). These are a typical mechanism whereby a company solicits for bids/proposals by other companies to take on work under a particular potential contract. These RFPs, which are essentially large disclosure exercises combined with a commercial pitch, are often heavily weighted by volume toward box-ticking. So, while the sections of the RFPs that might win the deal for a

---

[22] Ferraro, M. *et al.* (2023) *Ten Legal and business risks of chatbots and generative AI.* www.techpolicy.press/ten-legal-and-business-risks-of-chatbots-and-generative-ai/.

prospective vendor is in the commercial proposal, a deal can be lost before it has legs should the box-ticking fail. Examples include whether a vendor has specific insurances, meets particular legal requirements, conforms to particular accounting standards, or employs personnel in particular regions.

While skilled professionals in company legal, technical, and finance teams may hold this information, it is commonly only updated periodically. Therefore, capable of being delivered again and again within specific time frames without change. Even when there is change (e.g., the legal compliance of a company has been audited annually), updating this information is often straightforward and delivery of that information even simpler. Before the use of chatbots became commonplace, or at least rapidly adopted, companies often either

1. Relied on those who guarded the knowledge to repeat the data entry again and again or

2. Had a self-service simple document or spreadsheet with pre-determined answers which relied on the individual using it to know what they were looking for

AI-powered chatbots, however, can understand a multitude of natural language queries and deliver answers in a practical and concise way. What this means is that the key teams within a company (e.g., legal) can keep the data up-to-date, from which the chatbot draws. Then, the commercial teams can use that information to swiftly populate the RFP that they are the commercially minded to swiftly complete. This frees up specialist teams such as legal and accounting to focus on higher-complexity challenges while reducing commercial bottlenecks. Some chatbots/AI tools are even capable of auto-populating the RFP. However, these are typically more specialized and require greater setup and financial investment. Other uses of AI-powered chatbots for legal and accounting professionals supporting businesses are varied and new useful applications are frequently arising. For example, Danielle Benecke (the head of machine-learning practice at law firm Baker McKenzie) believes that AI tools are "especially useful for handling the legal fallout from common issues like cybersecurity incidents."[23]

What AI-powered chatbots such as these clearly illustrate is that there is a lot of low-risk low-hanging fruit for professionals to provide their services, particularly when they are delivered within a company rather than to an external customer. Overall, it is how useful a tool is (any tool, whether powered by AI or not) and the outputs of it to an end user/end client that will determine the satisfaction it delivers and therefore the volume, consistency, and quality of usage. Typically, tools that try to "boil the ocean" by marketing themselves as a panacea with the ability to solve all problems, may in reality be those least suitable to deliver real measurable value to the daily work of a professional for their end client.

Recall principle 1 (*If you are regulated – it's your regulatory responsibility*). With this in mind, it makes more sense, most of the time, for professionals to focus on using AI tools to help with the delivery of the lowest risk and lowest value, but often heavily time-consuming, aspects of their client service. Therefore, a key to success for professionals utilizing AI tools to deliver good service to their stakeholders is identifying what is most valuable to the company (whether that is an internal or external client) and effectively achieving the least skilled aspects of the professional service via the most useful mechanisms.

Of course, this will not be possible in all circumstances, but the volume is so significant in many organizations and for many lawyers that some think the use of AI systems in law may be the pinnacle of AI application

---

[23] Benecke, D. in Smith, M.S. (2024). *AI could become a lawyer's greatest help in the courtroom.* [online] Business Insider. Available at: www.businessinsider.com/artificial-intelligence-use-in-law-firms-legal-cases-2024-6 [Accessed 20 Jun. 2024].

due to its ability to reduce or eliminate tedious tasks.[24] While it is likely that, in reality, more societally important use cases for AI will arise – it will certainly be industry-transforming across the professions. In medicine, it is very hard for a doctor to determine without some patient interface what the most useful efficiency of using an AI tool might be. But, looking beyond just doctors and into wider healthcare services, it is easy to see how AI-powered refilling of prescriptions, or AI-supported reminders for health check-ups could all be useful and make significant differences to the quality of patient outcomes. This may also be useful for freeing up medical practitioners, who are over-stretched in most countries, to focus on higher-skilled and more complex patient care.

If You Don't Know Where the Data Goes, or Who Can See It, You're Taking a Bigger Risk

Practitioners within the professions handle, analyze, and interpret some of the most important and sensitive personal information of their clients. It doesn't require explanation to easily understand that most people would care where their private medical information (for example) goes, where it is stored, who can see it, who can use it and whether it is reasonably protected. Financial and legal information can often be as sensitive or important as medical information, so the same principles broadly apply to lawyers and accountants as they do to doctors. While less regulated or less sensitive professions (such as engineering) still have some frameworks of confidentiality and requirements as to how customer data is stored, particularly when it is of a personal nature, risks are generally going to be lower than the three professions which are the focus of this chapter.

---

[24] Ziniti, C.. in Smith, M.S. (2024). *AI could become a lawyer's greatest help in the courtroom.* [online] Business Insider. Available at: www.businessinsider.com/ artificial-intelligence-use-in-law-firms-legal-cases-2024-6 [Accessed 20 Jun. 2024].

As a starting point, it's important to remember as a provider of legal, medical, or accounting services that no data is ever entirely secure.[25] There have been countless examples of data breaches which involved no AI whatsoever. Often, these breaches or leaks of personal or sensitive information are not as a result of technology at all, but as a result of human error,[26] of poor human judgment.[27] Even in cases of malicious intent, it is common that the leak or breach is not caused by inherent technological flaws, but instead as a result of human opportunism against poor human judgment through social engineering, phishing, or many other potential ways in which targeted breaches occur.[28] Real "hacks," where a technological weakness is exploited rather than a human weakness to access technology, are comparatively rare.[29]

---

[25] Read, C. (2022). *Can NHS data ever be considered truly secure?* [online] Health Service Journal. Available at: www.hsj.co.uk/technology-and-innovation/can-nhs-data-ever-be-considered-truly-secure/7033923.article. [Accessed 20 Jun. 2024]

[26] Sjouwerman, S. (2020) *Stanford Research: 88% of data breaches are caused by human error.* https://blog.knowbe4.com/88-percent-of-data-breaches-are-caused-by-human-error#:~:text=Researchers%20from%20Stanford%20University%20and,caused%20by%20an%20employee%20mistake. [Accessed 19 June 2024]

[27] Maurer, R. (2023) "Human error cited as top cause of data breaches," *SHRM*, 21 December. www.shrm.org/topics-tools/news/risk-management/human-error-cited-top-cause-data-breaches#:~:text=A%20new%20study%20reveals%20that,largest%20cause%20of%20security%20breaches.

[28] Cveticanin, N. (2023) DataProt, https://dataprot.net/statistics/hacking-statistics/.

[29] Watts, E. (2023) *9 Most common causes of data breaches | Data-Breach.com.* https://data-breach.com/9-most-common-causes-of-data-breaches.

Most economically developed countries have some form of data privacy laws,[30] the most famous being the GDPR/UKGDPR which govern the EU and the UK. Professionals are not exempt from the requirements of such privacy regulations to treat personal data of individuals with appropriate sensitivity and protection, while avoiding sharing that data with anyone or any company that does not have the individual's consent or without other legitimate grounds. So these privacy laws, in practice, sit as a foundation to how personal data can be handled. Often, the same frameworks are used by companies to manage how confidential data is generally handled, even if that data does not contain personal information. On top of this, each professional regulator will have their own guidelines for sensitivity, confidentiality, and overall duties of care regarding end clients and their data.

So, as a professional with a legal obligation and duty of care to an end client, it is critical that before using any new technology (irrespective of whether it is powered by AI) to have some reasonable level of understanding as to where the data is going. For example, when data is entered into an AI-powered chatbot such as Gemini or ChatGPT. When using AI systems that may interact with clients' personal data or sensitive data (e.g., patient medical history, or even simply their name and address), the key data privacy principles to bear in mind are

1. Whether the amount, detail, and sensitivity of the data is appropriate for the professional to gather, hold, and analyze. Or, is it excessive and can it be reduced?

---

[30] unctad.org. (n.d.). *Data Protection and Privacy Legislation Worldwide | UNCTAD*. [online] Available at: https://unctad.org/page/data-protection-and-privacy-legislation-worldwide#:~:text=137%20out%20of%20194%20 countries. [Accessed 19 June 2024]

2. After the data has been used for the primary
   purpose (e.g., to determine a diagnosis, or give legal
   advice), does it need to be retained. Or, can the data
   be protected by being deleted within an appropriate
   time frame?

3. Does the end client know what data is actually
   being held and processed by the professional, with
   sufficient understanding to either give continued
   and informed consent or to ask for that data to be
   deleted? Or, if consent is not explicitly given, is there
   another legitimate legal basis for the data processing
   to take place/continue?

4. Is the data being sufficiently and appropriately
   protected in every regard by the professional,
   whether that is through sufficient security or by
   minimizing the data gathered and held?

It is equally important to understand what is happening to that data
once it is within that AI provider's systems. Looking at the regular free-
to-use (non-paid subscription) version of ChatGPT as an example, as at 1
January 2024 (worth stating this date as things can change), OpenAI as the
provider of ChatGPT reserves the rights to collect and retain considerable
information on their users. This includes email addresses and IP location
as well as your physical location (if it can see that via your IP). More
importantly for a professional, it retains conversation history (the prompts
entered into the system. This may include information you use or enter in
your chat/prompts. Therefore, irrespective of how sensitive or personal
that information may be, it may be fully retained by OpenAI and used

for broadly any commercial purpose they wish.[31] Typically, for adding to their dataset for their AI model,[32] but it shouldn't be assumed that they are restricted to just that purpose. Crucially, OpenAI's terms and conditions state that they do not use the data of their paid-for account customers and for those who access the system via an API (a type of technological integration, potentially via a different tech platform or interface).

Applying the example of entering potentially sensitive personal data into ChatGPT – lawyers, doctors, or accountants wish to use AI systems as professional tools, the devil is in the detail. It is crucial that the practitioner has a strong understanding (or has relied on someone they trust to have that understanding which can be relied upon) of what exactly an AI system's provider is offering, and they take in exchange with respect to the storage and usage of data which is entered into the tool. In practice, this may simply mean reading terms and conditions (or at least summaries of those terms) and/or often using more premium versions of the tools which allow users to have greater protections and discretion as to the way the data is handled.

If in doubt, a professional should seek guidance from and place reliance on their employer. Unless the professional is acting as a sole practitioner (e.g., a one-person law firm or a single-doctor medical practice), their law firm, accountancy practice, or medical practice may have their own views or policies on the usage of AI tools. It may be a breach of such internal policies for a professional to use any AI tools, or enter client data into any third party systems whatsoever (whether AI powered or not) without the approval of their employer. In most cases, a

---

[31] Arnott, B. (2023) *Yes, ChatGPT saves your data. Here's how to keep it secure.* www.forcepoint.com/blog/insights/does-chatgpt-save-data#:~:text=ChatGPT%20collects%20both%20your%20account,use%20in%20your%20ChatGPT%20prompts.

[32] OpenAI. (n.d.). *How your data is used to improve model performance | OpenAI Help Center.* [online] help.openai.com. Available at: https://help.openai.com/en/articles/5722486-how-your-data-is-used-to-improve-model-performance. [Accessed 20 June 2024]

doctor would not be within the bounds of their professional duties or their contracts of employment should they use an AI tool to assist in diagnosis without the consent of the hospital or medical practice which employs them – whether or not any of the entered data is identifiable or attributable to a patient.

Principles As a Non-professional

Whether you are simply preparing your wills and using the services of a lawyer, or a patient preparing to undertake a medical procedure, if you are the end client of a professional then you can and should be in control of your decision-making. At the very least, if delegating decision-making authority to the advice and recommendation of a professional, it is worth having a clear understanding of the basis on which the advice is given.

Using a Self-Service Approach

Before considering how to be safe when using a professional who relies on or utilizes AI, it's worth remembering that there are a huge amount of publicly available and often free tools which can be utilized to obtain some valuable understanding or advice, without even involving a professional. Broadly, non-professionals can still use AI tools to better learn and understand specialist subject matter and for purposes that don't require the AI systems to give advice. Asking an AI system to provide scheduled reminders to take medication[33] (so long as the prescription and timing is determined by a medical professional), may be a low risk and very helpful use case which requires no AI-generated advice.

---

[33] Tapper, J. (2024) "Warning over use in UK of unregulated AI chatbots to create social care plans," *The Guardian*, 10 March. www.theguardian.com/technology/2024/mar/10/warning-over-use-in-uk-of-unregulated-ai-chatbots-to-create-social-care-plans.

Unfortunately, using AI for self-learning (just like Googling one's own symptoms) may create as many problems as it solves. For example, if an individual who is not medically trained or has only a rudimentary foundational understanding of human physiology is attempting to use an AI tool to self-diagnose – the results could be disastrous. This would be particularly bad if the individual then did not seek medical advice from a doctor when their condition is potentially time sensitive. Asking a system such as ChatGPT for any kind of advice relies on the data sets that ChatGPT's owners (OpenAI) have to deliver recommendations or predictions. The potential issues most prominently arise because ChatGPT and other similar chatbot-style systems (at least the public, non-specialized versions) do not specialize in collecting or analyzing medical data and evidence, even if it may contain large volumes of this type of information in an unrefined, unverified, and uncategorized form.

It is the equivalent of seeking a consultation from a person who can regurgitate impressive volumes of information, but cannot separate what is important from what isn't, has no practical medical experience, and has no particular interest in the context and no ethical/moral parameters. Training to be a professional (doctor, lawyer, accountant or otherwise) is more than just acquired knowledge – it is the culmination of experience which allows for the application of refined judgment and an understanding of ethical nuances in the provision of advice or a service.

That is a core reason why, in most economically developed nations, each of the professions has a period of supervised training, where the professional's competence is framed (and in some cases signed off) by a more experienced professional in the same field. In many situations, it is also worth remembering there is often a genuine benefit for all parties to direct end client to professional interface.

Returning to the medical example, an AI tool can tell you if a type of procedure may be typical for a condition, but it will be assuming the inputs describing your condition are right to begin with, rather than necessarily being critical of the parameters it is fed. AI will not, in reality, tell you if it *believes* a procedure is right for you with any real degree of empathy and nuance because it is not capable of belief. AI will struggle with the causal reasoning that a doctor with face-to-face (even virtually) can interrogate and apply.[34] Critically, AI systems cannot answer, with any real understanding of the ramifications, the most difficult questions where an end client/patient asks a professional "What should I do?"

If a non-regulated non-professional attempts to use AI tools to provide a regulated service, even with the best of intentions, this could have severe consequences. In many jurisdictions, this type of action is simply illegal. Providing advice to others in this way, particularly if the recipient of the advice has a reason to believe they should place genuine reliance on that advice, could result in the provider of that advice (the non-regulated individual, rather than the AI provider) facing personal legal liability.[35]

Using Professional Services

Using the services of a professional is lower risk, but typically higher cost than personally using AI tools to secure the outputs of professional services.

The following are key questions to ask and understand:

1. **Is the advice being given by a professional within the industry holding all of the qualifications that should be reasonably expected?**

---

[34] Richens, J.G., Lee, C.M. and Johri, S. (2020) "Improving the accuracy of medical diagnosis with causal machine learning," *Nature Communications*, 11(1). https://doi.org/10.1038/s41467-020-17419-7.

[35] Byrom, N. (2023) *Unregulated AI legal advice puts the public at risk.* www.lawgazette.co.uk/commentary-and-opinion/unregulated-ai-legal-advice-puts-the-public-at-risk/5118241.article.

For example, a US lawyer who has passed the relevant state bar exams. Taking it a step further, the professional should give advice within the scope of their expertise and qualifications. If a doctor, are they providing medical advice within the scope of their expertise? It would be unreasonable to expect an orthopedic surgeon to be capable (or even willing) to give prescriptive advice or action for a gastroenterological issue. If they are an accountant or lawyer trained and specialized in the UK, it would be unreasonable to expect them to be capable/willing to deliver advice regarding tax or the law in another country.

2. **Does the professional have the appropriate and typical insurances?**

In most professions, either the individual or their employer should carry significant insurance coverage to protect them and their end clients from unfortunate mistakes or even outright negligence. It is highly unlikely that a lawyer at a major firm won't have or be protected by some insurance coverage, but it is often worth enquiring and requesting confirmation. In many jurisdictions, it is considered negligent to carry out certain professional practices such as law or medicine without insurance coverage.[36]

---

[36] *What does "adequate and appropriate" insurance mean for a freelance solicitor?* (no date). www.lawsociety.org.uk/contact-or-visit-us/helplines/ practice-advice-service/q-and-as/what-does-adequate-and-appropriate- pii-mean-for-a-freelance-solicitor.

3. **Is the advice given on the basis of AI-generated recommendations?**

Given the focus of this book, this is the key "new" question to understand when taking advice from any regulated professional. Within the parameters of this question, it is useful to understand or at least to ask some additional questions. The reason it is often useful to ask these additional questions, even if an end client does not *need* all of the answers, is that it can give great insight as to whether the professional has given the key issues sufficient consideration. Accordingly, whether it is reasonable to continue to rely on their professional expertise when they are using AI tools to facilitate or potentially improve their practice.

Additional questions:

1. **How has the AI tool been selected?**

Was it just picked at random, on the basis of good marketing? Or, was it recommended as being the right tool for the job by some kind of objective, or reputable subjective, third party?

2. **Has the AI tool been reviewed, approved, or at least used by any other major or reputable body or organization within the relevant field?**

Does the regulator within the field recommend this AI tool? As AI is a rapidly developing field, the scale, quality, and specificity of AI tools evolve and the professional services provider may be negligent for using unapproved, untested, or insufficiently specialist tools.

3. **Does using the AI risk invalidating the relevant professional insurances (e.g., medical negligence insurance)?**

It's quite possible that should a professional place total or near-total reliance on an AI tool to do their job for them, they may be acting outside of the reasonable bounds of their insurance coverage. This may be even beyond the scope of what policy aimed at covering professional negligence. Blind delegation or unverified dependency on sources is typically frowned upon within all the professional services (quite rightly, as that's what the professional is being paid for), and insurers generally take a similar view.

In future, we may see more insurers specifically carve out reliance on AI systems from professional insurance coverage. For now, most insurance policies appear to take the same basic approach when a professional has relied on AI, as any other sole reliance (e.g., relying on a seriously out of date textbook), that is, that this may simply be poor practice and whether it is appropriate will depend on the facts of the scenario. Potentially, however, this approach could be considered below par to the extent that it falls below the typical "errors and omissions" that might be covered by an insurance policy aimed at typical professional protection.

Remember, insurance comes into play to protect the aggrieved end customer, but insurers don't exist for that purpose. In the context of professional liability, they exist to cover the reasonably foreseeable, potentially even quite bad, mistakes of a professional. Some insurances, for example, those designed to protect patients from medical malpractice by their doctors, may still pay out even when the doctor, nurse, or medical institution has fallen below even very low standards. However, as a user of professional services, it would be dangerous to automatically assume that because a professional is using AI, that any greater protection is afforded – should negligent advice be given. AI is just another tool. So, when asking your doctor/lawyer/accountant to confirm whether they use any AI tools to provide you with their service, it may also be worth asking them to confirm that the usage of those AI tools won't affect their professional insurances. Unless they are *reliant* on those tools, it likely shouldn't in many countries – if they have good insurance.

4. **Is the professional likely to be (now or in the future) negligent if they fail to use helpful AI tools, when they otherwise could/should?**

   It's worth remembering (as further discussed in the "Principles as a Practitioner" section) that AI is used for a reason. It can accelerate, improve accuracy, facilitate collaboration, and provide new ideas to its users. So, if your doctor, lawyer, or accountant is providing you with advice and they are failing to use AI tools which are, or become, their industry norm – that could be negligent in and of itself. The world isn't quite there yet, but we're not far.

To illustrate this future, it would already be likely that in today's world, an accountant could be considered as potentially negligent for a poor outcome, should they carry out complex calculations with a pen and paper rather than using cheap and readily available tools such as Microsoft Excel. At the very least, the accountant in this scenario would be expected to verify their initial calculations even if they have been insistent on an outdated process. It is not inappropriate to expect and ask a professional delivering a service to always use the most recent and useful tools, particularly if they are already mass-adopted and the norm within that profession.

Even in a near-future world, where professionals may be negligent if they unreasonably forgo the potential advantages of using AI tools, they do not owe an end client a duty or obligation to use any particular technology that the end client wishes. The reasonable discretion of professionals remains. Therefore, it may (likely) be the case in the near future that as a patient, it would be reasonable to expect a doctor to double-check a diagnosis with an AI-powered second opinion. However, that doctor will not have to select a particular brand or style of AI-powered tool based on the preference of their patient.

Key Considerations

1. **Duty of care**: "The AI made a mistake" is not sufficient justification for very poor or negligent service by professionals. Be aware and actively recognize that this really means "We made a mistake" and that AI was just a tool used in the process of service delivery. Professional service providers, whether it is the individuals or their employers (such as hospitals, law firms, and accountancy firms), owe the same duties of care to end clients, irrespective of whether you use AI or not.

2. **Insurance:** There is a real ethical and practical issue when a professional service provider has insufficient insurance, or insurance which may carve out AI usage from coverage. As the professional service provider is ultimately responsible to the end client and to the relevant regulator, it is their responsibility to maintain appropriate insurance types and coverage levels.

3. **Transparency:** It is going to become increasingly typical for end clients to be willing to ask their doctors, lawyers, accountants, and other professionals whether they have used AI in the provision of their services. Transparency of methodology is already a cornerstone of professional services, and the use of AI may become an expected disclosure. End clients should feel comfortable asking the question of whether AI has been used. Professionals should equip themselves to answer with clear rationale as to whether and why AI has or has not been used, and why.

4. **One size rarely fits all:** Tools that claim to be a panacea (i.e., a generalist solution to solve all of the problems faced by a professional, or to replicate all the capabilities of a professional) should be used cautiously. Generalist tools are often less useful or suited to complex or sensitive work as they lack the specialization, focus, and targeted context in their development and operation to be relied upon to a meaningful extent. However, these tools can still help professionals, or those self-serving and bypassing professionals, to gain significant speed advantages and a broad understanding of simple matters.

5. **Self-service**: Usage of AI-systems beyond self-education should be done with great caution. There are an increasing amount of self-service AI systems that non-professionals can use as tools to increase their own knowledge and understanding. The can help non-professionals get more out of any follow-on discussions (i.e., as end clients) with regulated professional service providers. Non-professionals should be mindful that although these AI tools are useful, fast, and often cheap – they are not experts who have had the benefit of direct human interface and situational context. Most are not designed or intended to be used in a way that serious (medical, legal, accounting) reliance should be placed on their outputs.

CHAPTER 8

Summary and AI for All

How to practically apply this book

To navigate the opportunities, risks, and legalities of AI, try to ingest the Key Considerations of each chapter. These aim to be practical and use-case-specific. In addition, these have been distilled as set out in the following summary so that you can follow a cheat-sheet of Key Considerations.

Summary of Key Considerations – How to Use AI Effectively and Safely

Consideration 1

If you work in a job that does not primarily require physical skills and exertion, AI systems will likely become important, common, and eventually integral to your day-to-day work. Eventually, physical labor will be supported by AI-powered robotics, but we're not there yet.

Consideration 2

Education regarding AI skills, and the use of AI systems more broadly in educational programmes, are lagging compared to the pace of real-world adoption of AI by students and in the workplace. Increasingly, employers will be expecting graduates into the workplace to be AI-native or to, at least, be competent in operating various AI systems.

Consideration 3

If you create content using AI systems, you risk not being able to prevent others from using that same content once you make it public. This means that the "IP rights" you might otherwise have or be able to secure for innovative or original creations (whether written, visual, audio, photographic, artistic, or commercial) might not be quite as readily available or obtainable as you might expect from an equivalent creation made without the use of AI.

Consideration 4

When using AI systems to create images, videos, or words, it will typically be the user/creator of the AI system who is responsible for the potential harms that can arise. These harms can result in civil or even criminal legal consequences. If the content generated by AI systems infringes/breaches the IP rights, privacy rights, or confidentiality of third parties, the primary risk rests with the user of the system. Users should not expect to have recourse or protection from the AI system provider, even if the AI system provider faces its own, separate, legal consequences.

Consideration 5

If data is put into AI systems by users, it is largely the case that the provider of the AI system can/will use that data unless they very clearly say otherwise. Once the data is in, it's very hard to say with certainty where it will end up and how it will be used. So, when you use an AI system, the more you control and consider what data you put into it, the less likely you are to later have regrets. Balance this against the potential practical upsides you might gain from using the AI system.

Consideration 6

Misinformation and fraud can be facilitated by the involvement of AI systems, as with many new technologies. Checking sources and the validity of the information and media you consume has never been more important.

Consideration 7

AI systems and algorithms probably affect your life far more than you realize (at work and in your personal life), even when they are just filtering your potential choices rather than making them for you. As the sophistication of AI systems increases, the impacts on your day-to-day life are likely to grow. You can either lean in and understand how AI works and how to take steps to control the impacts and effects in positive ways, or simply try to reduce your reliance on technology. As the latter isn't likely to be an option for most people, learn whatever you can – read the news when new AI systems come out, play with them yourself when possible (particularly if they are free or cheap).

Consideration 8

AI systems can be powerful tools for education on financial, legal, medical, and other important matters. This can grant users empowerment and independence. However, AI systems don't protect a user from making poor or unlucky financial, legal, medical, or other high importance/high risk decisions. They also are not typically regulated, and therefore don't offer any protections should an important decision go wrong (e.g., choice of investment, treatment, or a bad legal arrangement). So, unless you are extremely experienced – you should exercise great caution. For example, when using AI with respect to finances, only invest/risk an amount you can afford to lose and try not to invest or make financial decisions using AI systems without considering or seeking a second opinion from a regulated professional or organization.

Consideration 9

AI cannot replace the heart and soul of real human-to-human interaction, nor the comfort and sub-textual understanding that it can bring. So, we shouldn't expect AI to successfully fully replace important roles within the "professions" such as law, medicine, accountancy, or other roles where human understanding is a key element – such as social work, nursing, or therapy. Without human involvement, in many circumstances, the risk of important details or context being missed is likely to increase which may in turn lead to worse outcomes. Importantly, for when we are at the receiving end of these important services, we should actively *want* to preserve the humanity within these roles.

Consideration 10

The law on AI is moving fast, with a focus on protecting ordinary people from the potential harms of AI when misused by other individuals or commercial interests. Nevertheless, the law moves slower than people and companies. So, depending on the country in which you live, you can expect some *basic* protections and safeguards from the law – but there is no replacement for practical safety (i.e., the preceding Considerations 1–11). There is a big difference between laws existing and actually being enforced. So, focus on using AI systems for where you have identified there will be a truly valuable upside which is worth any potential risks.

The Future

For the ordinary user of AI, the potential use cases are not confined to those already discussed in this book. Existing applications of AI in almost every field are nascent, at best. It is impossible, with any reasonable degree of certainty, to identify what the daily realities of AI's impact on people's lives will be. Five to ten years from now, the ways that ordinary people (not just advanced users or early adopters) interact or are affected by AI may be dramatically different, we just don't yet know how with real certainty.

Even referring to AI as a single thing, like we may do with the Internet, may become old-hat. Already, we differentiate between aspects of the Internet – "streaming," "apps," "in-browser" being just a few. On the other hand, it is increasingly likely that various types of AI (e.g., those relating to image vs. text or analytical vs. generative) may become a single seamless entity – as we are already seeing with early cross-platform systems such as ChatGPT 4o, which have multi-format capabilities across text, images, audio, and video.

Companies in the AI space face an oncoming storm of uncertainty with a need to adapt because of rapidly developing AI laws and regulations. The EU AI Act 2024 increases the onus on AI system providers to be more transparent about their data collection, data usage, and output accuracy. In addition, there are new legally enforceable limitations or outright prohibitions on some of the potential real-world applications of AI.[1]

In early 2022, there was a surge of predictions that, within two2 to five years, we would be living in an AI-revolutionized world which would be almost unrecognizable.[2] Fast-forward, and we know that a lot of this was hyperbolic enthusiasm. While the thesis that AI changes almost everything might be right, it is clear that the legal, ethical, and practical principles that govern the lives of individuals and business remain fundamentally the same. For now, AI is an accelerator rather than a revolutionizer. While this is potentially disappointing or uninspiring, that is, just doing what we already do but faster or more easily, it is very hard to predict revolutionary developments and applications of AI that may arise in future. In the early 2000s, with the Internet already over a decade old, few would have accurately predicted the interaction between mobile devices, social media, and people's ordinary lives as well as the wider economic effects.[3] In its infancy, early usage of the Internet was strikingly similar to AI – an accelerator of existing processes and democratizer of information, but

---

[1] Braun, M., Vallery, A. and Benizri, I. (2024) "Prohibited AI Practices—A Deep Dive into Article 5 of the European Union's AI Act," *WilmerHale*, 8 April. www.wilmerhale.com/en/insights/blogs/wilmerhale-privacy-and-cybersecurity-law/20240408-prohibited-ai-practices-a-deep-dive-into-article-5-of-the-european-unions-ai-act.

[2] Gil Press (2022) "What happened to AI in 2022?," *Forbes*, 30 December. www.forbes.com/sites/gilpress/2022/12/30/what-happened-to-ai-in-2022/.

[3] Amelia Tait, The New Statesman, www.newstatesman.com/science-tech/2016/08/25-years-here-are-worst-ever-predictions-about-internet.

without truly original applications or benefits. Commercialization and sustained usage by a mix of individuals and businesses led to genuinely innovative uses and innovations.

Lack of predictability makes appropriate regulation of any technology almost impossible, unless new laws are created with inbuilt expectations of, and actual mechanisms for, change. The law will need to adapt over time as the relevant technology advances,[4] irrespective of current unpredictability. We can already see that uncertainty relating to how AI technology (and AI uses) will look in the near future are an impediment to consensus of appropriate regulatory frameworks. Even the first major national/supranational legislation within the European Union, the EU AI Act 2024, will inevitably be out-of-date in short order, as AI adoption accelerates, unforeseen use cases will arise. So, it is important to look at the past to understand how things are likely to unfold in reality.

Case Studies

With these key considerations in mind, an individual or business must still navigate uncertainty to capitalize on new technologies without being exposed to disproportionate risk. This applies whether the desired outcome is an improvement to personal lives, careers, or commercial outcomes. As it is not possible to predict the future, the only data we have is historic information and analysis on successes and failures in other tech booms. This means that there is real value to reviewing aggregated information and/or specific examples to glean where people got it wrong, and where they got it very right.

---

[4] Eggers, W.D., Turley, Pankaj, M., Kishnani, K. (2018) Deloitte "The Future of Regulation, Principles for regulating emerging technologies." www2.deloitte.com/us/en/insights/industry/public-sector/future-of-regulation/regulating-emerging-technology.html.

Even when looking at specific case studies, the goal is not to exactly replicate or avoid past events. Technologies are never exactly like-for-like, so that would likely be ineffective. Instead, when reviewing these case studies, the point is to learn from the facts and use those as qualitative illustrations of key considerations. Then, to concisely consider and how those give rise to specific lessons (see the end of each case study). Ideally, this helps demonstrate the practical strategies which can be applied by almost anyone, with lessons that are actionable rather than abstract.

Social Networks and the Data Boom

The origins

Social media and the current Internet's love of data-for-services (user data for commercial services) are one of the great stories of the Internet age and also directly contributed to some of the most powerful multi-jurisdictional legal guardrails to secure the privacy of consumers. On the February 4, 2004, TheFacebook was launched. In hindsight, this is arguably *the* archetypal example of an unpredictable, unprecedented, and hugely profitable future stemming from a new technological idea. The idea of a digital social network was hard for many people who were not the initial adopters (i.e., students) to understand, even with some other community-based systems already gaining significant traction. The benefits did not seem clear to many users, other than simply seeking human connection via a digital medium. The concept of putting large amounts of personal data for the world to see, even just a photo and information about where a person went to school, was odd to most people other than students – particularly as even one generation older were unlikely to be Internet-native in their own education. Nevertheless, these reservations did not last long, with billions of people eventually becoming daily users.[5]

It took Meta (previously TheFacebook, then Facebook) over five years to turn a profit.[6] Facebook figured out and launched its advertiser-friendly interface in 2007 while it benefitted from an enormous growth in the advertising technology and online search sectors. This, combined with marketers within companies and agencies looking for new ways to target consumers with effective advertising, was a recipe for huge revenue that was not immediately obvious at the company's inception. It was game-changing. In combination with the commercial diversification, Meta developed what Gil Press of Forbes called an "advertising-depended business model based on collecting and analysing users' data, content and actions."[7] What is most important to recognize is that this was not the initial business plan or intended use of user data by Facebook – the plan came later, and the users had already signed up and happily handed over significant data to enjoy the outcomes of that data exchange. For example, a user showing off to a friend that they were now "in a relationship" or were popular by virtue of their connections.

Shifting to a monetized data digital society

It was the focus on data collection and exploitation (in the literal sense of the word, rather implying this is inherently negative) which allowed Meta to grow while many of its initial competitors died. Though we may have fond memories of MySpace and Bebo, they are long-dead for the

---

[5] Barr, S. (2018) "When did Facebook start? The story behind a company that took over the world | The Independent," *The Independent*, 23 August. www.independent.co.uk/tech/facebook-when-started-how-mark-zuckerberg-history-harvard-eduardo-saverin-a8505151.html.

[6] Thompson, D. (2009) "Facebook turns a profit, users hits 300 million," *The Atlantic*, 17 September. www.theatlantic.com/business/archive/2009/09/facebook-turns-a-profit-users-hits-300-million/26721/.

[7] Gil Press (2022) "What happened to AI in 2022?," *Forbes*, 30 December. www.forbes.com/sites/gilpress/2022/12/30/what-happened-to-ai-in-2022/.

principal reason that they were quick to gather users but slow to figure out the pivot in their commercial model. These kinds of pivots are common among technology companies that succeed in the long term and should be expected by users, that is, on day 1, a company provides X service, on day 500 it provides X, Y, and Z services and the original X service may look quite different (often seamlessly, without the user even being particularly aware or unhappy with the change in dynamic).

While these commercial pivots from the founding business ideas are typically uncontroversial from a commercial perspective, companies still have to operate within the legal frameworks that place restrictions and limitations on their commercial options and drive. TheFacebook simply encouraged use and gathered data, long before it became Facebook and the current behemoth of Meta. The business model which expanded into massively more complex technologies and revenue generating mechanisms, came later and was based on the available resources which were already gathered. However, in 2017, the European Union's General Data Protection Regulation became law, the first major and impactful privacy law of its kind. Previous regulations/laws did exist but were either relatively toothless, rarely enforced, or became more understood and important as a result of GDPR's prominence. There are now numerous, and growing, privacy laws and regulatory regimes around the world. Meta is now so large, that inevitable complexities and problems arise with how data is handled and used. At the same time, Meta (like many other large technology companies) can often afford to simply bear the consequences if they are found to not be following the strict letter of the law[8] by a regulator or court. Much of the difficultly arises due to internationally conflicting approaches, with a more business-friendly and less consumer privacy-focused approach in the Unites States. Arguably, Meta and other large tech companies must continue to risk regulatory questions or infractions to stay at the cutting edge of innovation. Targeted advertising is essential to the

---

[8] Torchio, G. (2023) *Meta's "Pay or Okay": Is this the final challenge for EU GDPR?* https://epc.eu/en/Publications/Metas-Pay-or-Okay-Is-this-the-final-challenge-for-EU-GDPR~5672dc.

economic model of the Internet as a whole, with Meta being a pioneer rather than "the bad guy." If anything, they (along with a few other large players, notably Google) proved the business model for much of the current Internet. The alternative would broadly be that advertising-funded services, which are free at the point of the end user, become paid-for services. Anyone who even takes the briefest moment to consider how many websites they visit in a week, that they don't pay for, will suddenly realize that targeted advertising is the actual business model of most of the daily-use Internet.[9]

The ship has sailed

The core business model of all social media providers is now dependent on gargantuan quantities of data which flows into targeted advertising. So, regulatory frameworks to minimize and secure individual rights in data create challenges for social media companies and Internet-based technology companies in their ordinary commercial-scale data gathering and usage. The more data that large social media companies have, the greater the possibility for refinement of ever-increasingly targeted adverts. Some of these challenges are primarily practical rather than inherently unethical or nefarious. For example, for large international tech companies, it is often a practical necessity to transfer data between jurisdictions despite various legal and contractual barriers on where data can reside or be sent without the explicit written consent of the data subject (i.e., the individuals whose data is gathered and processed). Lawmakers and regulatory authorities broadly understand this. Nevertheless, they have acted against big technology companies with significant force. In 2023 alone, Meta was fined approximately €1.5bn by European regulators[10] for how and where it sends user data. Such fines are by no means unique to Meta, or even social media companies and are

---

[9] Waters, R. (2017) Financial Times "Facebook needs to make money by making lives better." www.ft.com/content/832a33ee-e0ed-11e7-8f9f-de1c2175f5ce.
[10] Bodoni, S. (2023) Bloomberg "Are you a robot?" (2023). www.bloomberg.com/news/articles/2023-05-22/meta-fined-record-1-3-billion-in-eu-over-us-data-transfers.

common for a range of practical and organizational issues on how data is gathered, stored, used, and erased. Many outside the UK and EU would argue that the practical risks to individual privacy were minimal as a result of these data transfers, so the fines are disproportionate - even if UK and EU regulations have potentially been technically breached. These are some of the biggest practical challenges that many large tech companies face in the real world of data dependency.[11]

Despite major regulatory battles, large technology companies which run on data collection remain highly profitable and the underlying business models are broadly unchanged – because people like their services and companies like Meta provide real consumer value. Therefore, despite being increasingly robust, the law has moved too slowly to be optimally effective and to recognize that a vast amount of technology users are willing to simply accept loose control of their own data in exchange for "free" services. Regulatory enforcement, including fines, may become more sizeable and more frequent as public opinion becomes more vocal on individual privacy. However, billions of people continue to use Meta's services – Facebook, WhatsApp, Instagram (among others) because they are entrenched into those ecosystems of communication and social interaction, while companies such as Meta have little alternative to gather, process, use, and sell the data of those users. It should be fairly assumed that companies such as Meta try to do so within the confines of what regulations broadly allow (it wouldn't be sensible to assume by default they would willingly make rods for their own backs). Of course, *how* that data is handled can always be improved, but fundamentally, the services are valued by their users for providing a powerful service – so they continue to be popular. This is something the law can do little about. Quite simply, many large data-dependent tech companies are now "too

---

[11] *Spain: AEPD fines Google €10M for unlawful transfer of personal data* (2022). www.dataguidance.com/news/spain-aepd-fines-google-10m-unlawful-transfer-personal.

big to fail" to borrow a phrase from Andrew Sorkin's somewhat equivalent analysis of the banking industry.[12]

Data into the AI age

Though not necessarily obvious in its formative years, the ultimate resource of the Internet age was data. Even more so in the social media age, and further still in the new age of AI. Combined with the hardware and energy materials required to power AI, data has transformed from a relatively simple commodity to be packaged and sold. It is now the linchpin material for technological development in AI systems.

The growth of large technology companies which are dependent on the data of their users have grown in ways that would have been difficult to predict in their early years. This teaches us that once users' data is in a system, it's impossible to be totally sure of where it goes, who will use it, or how. Meta and Google are prime examples of technology companies, among many, which have grown into conglomerates providing services and benefitting from diverse revenue streams. Many of these revenue streams are built on data, with new commercial applications for that data being constantly explored. For example, a photo posted via a social media application is now potentially usable in a wide variety of ways (depending on the social media company), likely far beyond what the poster of that image intended or understood. Such an image could even be used as part of the training set for AI systems.[13] This is not always inherently negative, but might fairly be considered to be a general weakening or potential erosion of individual legal rights and protections. And yet, billions of users

---

[12] Sorkin, A. (2009) *Too big to fail.* Viking Press.

[13] Carroll, M. (2024) *Meta is planning to use your Facebook and Instagram posts to train AI - and not everyone can opt out.* https://news.sky.com/story/meta-is-planning-to-use-your-facebook-and-instagram-posts-to-train-ai-and-not-everyone-can-opt-out-13158655#:~:text=Facebook%20and%20Instagram%20users%20in,used%20to%20train%20artificial%20intelligence.&text=If%20you%20have%20an%20Instagram,train%20artificial%20intelligence%20(AI).

are quite willing to put large amounts of their personal and professional lives into AI chatbots – often unaware or uninterested in the consequences.

So, users of AI systems should be selective and conscious about what data they are willing to give away and to whom. This mindfulness should be based on an understanding that the goalposts of the relationship between the user and the provider can change. The framework of the relationship, including how users pay for services or those services are otherwise monetized (i.e., they don't pay but someone else does because of their use) can change rapidly. In many cases, this change can happen without obvious or explicit consent of users.

Reliance on the law by ordinary users of AI technology for protection and practical guidance is optimistic. If potential reliance on legal frameworks to automatically protect users has any worth, it is only a long-term safeguard rather than something which currently provides much practical comfort. The law, by design and default, will always be slow to catch up to new technologies, and regulators are typically slow to enforce against technology providers who breach the trust of their users. The probable reality for users of publicly available AI systems (though unfortunate) is that the data given to AI companies may be kept or used beyond the original intention and without the originally agreed protections. In many cases, that data will be the data already available online as a continued legacy of over-sharing during the formative and continuing social media era. It is worth stressing that many professional-grade AI systems, typically more specialized systems, handle the data they receive with care, transparency, and diligence.

As a user of AI systems, be conscious to limit the data given to AI companies wherever reasonably possible. Once data is given by a user, it is in reality very difficult to erase the relevant digital footprint or reclaim the data, no matter how strong individual rights may be in theory. This is particularly important with sensitive or financial information, but it is difficult to predict how even relatively mundane information may be useful to a technology company – or even useful to be packaged and sold as a commodity to another party.

The simple lessons:

1. **Data minimization**

   To reduce the probability of potentially negative or harmful personal outcomes, this could be as simple as

   a. Avoiding putting sensitive information in an AI chatbot

   b. Unticking non-essential consents when signing up to a new tool or using an online service

   c. Limiting the information put onto social media to the information that you are willing/indifferent to being potentially leaked, retained, or used in a way outside of your initial intention

2. **Expectation management**

   a. When using an AI tool, don't expect that the content, functionality or usage of data will remain constant over time.

   b. Don't rely on the law to provide individual protection to users in a meaningful way. Even where there are legal consequences to large tech companies for breaches of laws and regulations, there is often very little direct recourse for the "victims," that is, the ordinary users.

   c. AI systems rely on data. So, don't expect that your data is not being used, or only being used how you have consented. In practice, AI system providers may place great value on the commercial need to collect and use good quality data, such that they are willing to take legal risks.

Amazon's One-Click (Investors)

September 1999 saw the granting of one of the most controversial patents and trademarks in modern times, Amazon's "One-click" online purchasing method. "One-click" was unique as a system that many of us now use daily, that is, the ability to purchase items through an e-commerce portal with a single click, having preset billing and shipping details. At the time, it was revolutionary. Patenting it was equally revolutionary.

The controversy of this patent centered about it being a stifle to innovation. While that is debatable, Amazon did gain significant advantages beyond simply a better product than its competitors. In essence, Amazon had cornered a method of interaction between a provider of digital services (or physical services purchased digitally) which was notably more efficient and facilitated growth using previously untapped shortcuts to efficiency.[14]

What this meant was that Amazon was able to use the legal frameworks of intellectual property registration and financial exploitation (in the literally, rather than pejorative sense) to great advantage, all facilitated by inventor-friendly legal frameworks to protect intellectual property. Arguably, this is exactly what IP registration rights (such as those discussed throughout this book) are designed to achieve.

On the other hand, investors in e-commerce businesses other than Amazon, particularly when Amazon capitalized on "One-click" to open its MarketPlace, had limited choices. Either

---

[14] Kartik Hosanagar, Leonard Lodish, Ron Berman, Knowledge at Wharton Staff, Knowledge. Wharton (UPenn), https://knowledge.wharton.upenn.edu/ podcast/knowledge-at-wharton-podcast/amazons-1-click-goes-off-patent/.

1. Accept that their investee businesses had a long-term commercial disadvantage. Many simply did accept this and either survived if they had sufficient resources (but didn't boom to the same extent as Amazon) or disappeared over time.

   Or

2. Accept that "One-click" was too useful and essential to growth, and they would have to pay whatever it cost to license it from Amazon.

Apple, on the verge of its own major growth explosion, saw the writing on the wall and chose option 2. Apple simply chose to pay Amazon in order to license the technology for its own benefit in music distribution on iTunes.[15]

On the other hand, Barnes & Noble tried to launch their own version of "One-click" called "Express Lane." Unlike Apple, Barnes & Noble seemingly believed that methodology innovation tied to technological innovation was harder to patent and enforce that pure technological ingenuity. When Amazon sued Barnes & Noble in October 1999 (just one month after filing the "One-click" patent), the effectiveness of being first to patent a genuinely useful innovation was clear. The judge issued an injunction against Barnes & Noble, effectively cutting off their ability to use their "Express Lane" in a single click fashion. It may seem that having to use two clicks instead of one wouldn't matter – but in the long term, it meant the difference of millions or even billions of dollars of consumer engagement. Further still, that injunction didn't prevent Amazon continuing to pursue and enforce against Barnes & Noble. The case was settled out of court in 2002, with the undisclosed settlement demonstrating Amazon's overwhelming ability to enforce its legal rights over IP.

---

[15] Apple Press Release, www.apple.com/newsroom/2000/09/18Apple-Licenses-Amazon-com-1-Click-Patent-and-Trademark/.

From an investor perspective, it is particularly interesting that at the time Amazon was patenting its technology, many other e-commerce businesses were receiving investment or being bought at astonishingly high valuations – many of which would collapse as a direct result of Amazon's ability to monopolize a method of business-to-consumer interaction so effectively. It isn't hard to imagine one or more AI companies figuring out similarly lucrative methods of monetizing their services or tools, using intellectual property law as boon to that growth.[16] So, for investors looking into AI companies, it's worth considering the method and ease in which users interact with AI tools and systems rather than just the usefulness of the tool in isolation.

Exuberant valuations based on technological hype could be the topic of an entirely separate book, but serve as a useful canary in the mine to indicate that specific legal challenges (which overlap or affect commercial issues) are also present. These legal challenges center around the intertwining of

1.   Intellectual property: Validity and certainty

2.   Regulation: Uncertainty and inconsistency

In the case of Amazon, it wasn't until 2017 that the "One-click" patent expired, by which time Amazon (having also re-invested and diversified) had become a company worth US$544 billion,[17] in some significant part due to its ability to streamline and dominate the online consumer shopping experience. The law never really shifted or acted to usurp Amazon's dominance in the way that many observers (or hopeful investors in other companies) may have hoped. Even regulatory regimes,

---

[16] PatentYogi.com, https://patentyogi.com/latest-patents/amazon/amazon-1-click-patent-was-worth-billions/.

[17] Wayne Duggan, US News, https://money.usnews.com/investing/stock-market-news/articles/2017-11-14/amazon-com-inc-amzn-valuation-stock.

such as those based on anti-trust (fair competition) principles, or data privacy (Amazon was/is holding large amounts of customer data in order to facilitate "One-click"), proved to be slow to react to this innovation. By the time data privacy laws, for example, were materially updated and impacted how Amazon could handle user data, over 15 years had passed.

It could be said that all bursts of macro-economic growth that arise out of a disruptive technology share commonalities. The dot.com bubble of the late 1990s and early 2000s, however, has incredible similarities to the AI boom of 2020 onwards. This is due to the proximity between the types of technologies, as well as the pace at which investor, provider, and user interest exploded.

For investors, this highlights the importance of assessing the IP landscape surrounding a new technology such as AI. Apple's decision to license the "One-click" feature underscores the strategic imperative of navigating complex legal systems in competitive landscapes dominated by pioneering or highly resourced incumbents. As AI is in an early developmental stage, we can already see exceptional financial valuations for business which, on paper, may be easily rocked by another company developing a product or shifting the norm of how people interact or use AI.

Equally, changes in regulation may destabilize the basis on which many AI companies are founded and scaling. As discussed in this book, numerous jurisdictions are producing their own sets of competing, and sometimes contradictory, AI laws to layer on top of their mixed data privacy laws and IP laws. We can assume that new laws (such as the EU AI Act 2024's application to banned and high-risk AI systems) may render some AI companies which have already received funding to be obsolete or incapable of delivering within the bounds of the law. Therefore, some investors will already have lost potentially significant amounts, and this is likely to grow and regional and international legal regimes develop and are enforced.

The simple lessons:

1. **The law can be a sword and shield.**

   Legal rights and protections, such as those regarding Intellectual property, may be weaponized in emerging tech booms. Regulatory guardrails can either facilitate or block those IP advantages or limitations. As a result, sudden turns in the tide of competition can occur which heavily favor a small group of entities (or even a single company) who have either been first-movers or are heavily resourced. Strategic advantages with real commercial outcomes can be rooted in effective use and navigation of legal and regulatory frameworks.

2. **The law affects real investing.**

   Valuations may be sky high for AI companies in the same way that they were in the early days of the dot.com bubble. Unlike many of the companies that failed in the dot.com bubble (and unlike Amazon), many of the AI technologies are real with legitimate and differentiated use cases. Not all of these companies will thrive or survive for reasons that are wider and more complex than the law alone. Nevertheless, shifts in the law may rapidly affect the viability of certain AI products, and therefore the validity of their companies making those products as prospects for external investment.

My Own Health (Professionals)

As a teenager in 2001, I suffered from serious spinal deterioration. Diagnosis took nearly two years, with treatment taking a further two years of surgeries and therapies. At several points, it was a very realistic probability that I would remain permanently unable to walk or in chronic pain. Doctors were often slow, inaccurate, inefficient during this process. Amongst an endless stream of medical professionals, my parents and I were met with a frustrating mix of false pessimism and unwarranted optimism based on misdiagnosis.

As already discussed in this book, AI may quite realistically replace many of the tasks currently carried out by medical professionals.

In 2016, Richard Susskind KC and Daniel Susskind published their analysis in the Harvard Business Review of the possible future role of doctors in the coming AI age. In their summary thoughts they identified two key issues with the perception by medical professionals that they may be "immune" to replacement by technology:

> *The first problem with this position is empirical. As our research shows, when professional work is broken down into component parts, many of the tasks involved turn out to be routine and process-based. They do not in fact call for judgment, creativity, or empathy.*

> *The second problem is conceptual. Insistence that the outcomes of professional advisers can only be achieved by sentient beings who are creative and empathetic usually rests on what we call the "AI fallacy" — the view that the only way to get machines to outperform the best human professionals will be to copy the way that these professionals work. The error here is not recognizing that human professionals are already being outgunned by a combination of brute processing power, big data, and remarkable algorithms. These systems do not replicate human reasoning and thinking. When systems beat the best humans at difficult games, when they predict the likely*

decisions of courts more accurately than lawyers, or when the probable outcomes of epidemics can be better gauged on the strength of past medical data than on medical science, we are witnessing the work of high-performing, unthinking machines.

Our inclination is to be sympathetic to this transformative use of technology, not least because today's professions, as currently organized, are creaking. They are increasingly unaffordable, opaque, and inefficient, and they fail to deliver value evenly across our communities. In most advanced economies, there is concern about the spiralling costs of health care, the lack of access to justice, the inadequacy of current educational systems, and the failure of auditors to recognize and stop various financial scandals. The professions need to change. Technology may force them to.[18]

In an earlier article by Daniel Susskind, he suggests the use of online automated tax software and well-known legal automation providers to a possible future where "increasingly capable systems will entirely replace the work of traditional professionals."[19]

However, another theory of his (on which he expands in the excellent "Future of the Professions" co-written with Richard Susskind) is more nuanced. He proposes that the daily accelerations and quality/consistency improvements in professions seem probable in the long-term, that is, that professions will look like "... a more efficient version of what we have today. Professionals use new systems to help them work in the traditional way..."

---

[18] Richard Susskind KC, Daniel Susskind, Harvard Business Review, https://hbr.org/2016/10/robots-will-replace-doctors-lawyers-and-other-professionals.

[19] Daniel Susskind, The Guardian, www.theguardian.com/commentisfree/2015/nov/02/robot-doctors-lawyers-professions-embrace-change-machines.

From a purely legal perspective, this is the only possible future based on how the law currently operates. For most legally developed jurisdictions, it is not a likely reality that the whole role of professionals is entirely replaced.

The emotional complexity and human contexts of medical matters (in particular, by comparison to other professions) means that machine-only systems may not only be less useful, but actually carry significant legal risk. Essentially, as long as the reasonable standard of care for a medical patient, or potentially any regulated professional, includes some of the capabilities that humans possess and machines lack, then the law requires those professionals to maintain their real skills and appropriate levels of patient interface. What is "appropriate" or typical may change over time. This is the case agnostic of whether the support/replacement of the professional is hardware or software, partially or entirely AI-driven.

The simple lessons:

1. **There is no perfect analysis. There are no perfect answers.**

   The analysis and advice of professionals is fallible. The analysis and advice of technology may, in some cases, be more empirically accurate but remains fallible.

   Users of AI tools for professional advice should generally consider gathering second opinions, even if that is just from an alternate AI tool, if possible.

2. **Human interface is hard/impossible to beat for some use cases.**

   In interactions, allow end users of professional services (such as medical patients) to provide context in an environment that may allow for context and risk analysis that technology cannot always match or beat.

213

3. **A human touch can soften a blow.**

    Some news is simply easier to receive and process when it comes from another person who can look the recipient in the eye. Not all outcomes are best served by simply being more efficient or accurate.

    The optimal positive outcome in many sensitive situations (whether medical, legal, or otherwise) can simply be delivering difficult news or advice in a way which makes it easier or more digestible for the recipient. This can directly improve the ability of the recipient to take action on the news, if possible.

4. **People hire professionals for their judgment, not the tools they use.**

    There remains room for professional judgment, which does not inherently exclude the value provided by AI systems. AI can support and reduce risks for professionals providing their judgment, rather than replacing the professional.

    To maximize the best outcome as the end user of professional services, it may be best to encourage and use (when possible) an approach which takes advantage of AI's empirical analysis combined with human judgment.

Index

A

© Harry Borovick 2024
H. Borovick, *AI and the Law*, https://doi.org/10.1007/979-8-8688-0400-7

GPSR Compliance

The European Union's (EU) General Product Safety Regulation (GPSR) is a set of rules that requires consumer products to be safe and our obligations to ensure this.

If you have any concerns about our products, you can contact us on ProductSafety@springernature.com

In case Publisher is established outside the EU, the EU authorized representative is:

Springer Nature Customer Service Center GmbH
Europaplatz 3
69115 Heidelberg, Germany

Batch number: 08698267

Printed by Printforce, the Netherlands